Philosophical Foundations of Quantum Field Theory

EDITED BY
Harvey R. Brown
AND
Rom Harré

CLARENDON PRESS · OXFORD

OXFORD
UNIVERSITY PRESS

Great Clarendon Street, Oxford OX2 6DP

Oxford University Press is a department of the University of Oxford.
It furthers the University's objective of excellence in research, scholarship,
and education by publishing worldwide in

Oxford New York

Athens Auckland Bangkok Bogotá Buenos Aires Cape Town
Chennai Dar es Salaam Delhi Florence Hong Kong Istanbul Karachi
Kolkata Kuala Lumpur Madrid Melbourne Mexico City Mumbai Nairobi
Paris São Paulo Shanghai Singapore Taipei Tokyo Toronto Warsaw

and associated companies in Berlin Ibadan

Oxford is a registered trade mark of Oxford University Press
in the UK and in certain other countries

Published in the United States
by Oxford University Press Inc., New York

Except where otherwise stated © Oxford University Press 1988

The moral rights of the author have been asserted

Database right Oxford University Press (maker)

Reprinted 2001

All rights reserved. No part of this publication may be reproduced,
stored in a retrieval system, or transmitted, in any form or by any means,
without the prior permission in writing of Oxford University Press,
or as expressly permitted by law, or under terms agreed with the appropriate
reprographics rights organization. Enquiries concerning reproduction
outside the scope of the above should be sent to the Rights Department,
Oxford University Press, at the address above

You must not circulate this book in any other binding or cover
and you must impose this same condition on any acquirer

ISBN 0-19-824289-1

Printed in Great Britain
on acid-free paper by
Bookcraft (Bath) Short Run Books
Midsomer Norton

CONTENTS

Notes on Contributors — vi
Introduction — 1

QUANTUM FIELD THEORY AS OBJECT OF PHILOSOPHICAL STUDY: TWO VIEWS

1. A Philosopher Looks at Quantum Field Theory — 9
   Michael Redhead
2. Foundational Problems in and Methodological Lessons from Quantum Field Theory — 25
   James T. Cushing

II THE PROBLEMS OF VIRTUAL PARTICLES AND RENORMALIZATION

3. Virtual Particles and the Interpretation of Quantum Field Theory — 43
   Robert Weingard
4. Parsing the Amplitudes — 59
   Rom Harré
5. Three Problems of Renormalization — 73
   Paul Teller

III COVARIANCE PRINCIPLES IN QUANTUM FIELD THEORY

6. Hyperplane-dependent Quantized Fields and Lorentz Invariance — 93
   Gordon N. Fleming
7. Gauge Theory and the Geometrization of Fundamental Physics — 117
   Tian-Yu Cao

IV MATHEMATICAL FOUNDATIONS OF QUANTUM FIELD THEORY

8. Why Should Anyone Want to Axiomatize Quantum Field Theory? — 137
   Ray F. Streater
9. The Algebraic Approach to Quantum Field Theory — 149
   Simon Saunders

Index — 187

NOTES ON CONTRIBUTORS

HARVEY R. BROWN, Wolfson College, Oxford

JAMES T. CUSHING, Department of Physics, University of Notre Dame, Indiana

GORDON N. FLEMING, Department of Physics, Pennsylvania State University, University Park, Pennsylvania

ROM HARRÉ, Linacre College, Oxford

MICHAEL REDHEAD, Department of History and Philosophy of Science, University of Cambridge.

SIMON SAUNDERS, Wolfson College, Oxford

RAY F. STREATER, Department of Mathematics, King's College, London

PAUL TELLER, Department of Philosophy, University of Illinois, Chicago

TIAN-YU CAO, Trinity College, Cambridge

ROBERT WEINGARD, Department of Philosophy, Rutgers University, New Jersey

Introduction

The conceptual foundations of quantum mechanics have been the object of intense interest and debate for both physicists and philosophers ever since the birth of the theory. Providing a coherent and comprehensive interpretation of the non-relativistic quantum algorithm has proved to be an ongoing challenge, and the vast literature on the subject is a testament to both the perennial fascination of the problem and to the absence of any durable consensus about how best to understand the theory.

If the foundations of non-relativistic quantum mechanics (QM) have received and continue to receive the serious attention of large numbers of natural philosophers, the same can hardly be said of the foundations of quantum field theory (QFT). Yet QFT, in its various versions, is widely regarded today as the most fundamental theory of physics, and has achieved in some well-known applications the most accurate, corroborated predictions in the history of physics. The rise in prominence of the theory in recent decades, following a period of decline after its foundations had been laid in the late 1920s and 1930s, has largely failed to wrest the principal focus of philosophical interest away from QM.

There are no doubt many reasons for the relative paucity of detailed investigations into the philosophy of QFT. One of them is perhaps a hasty adherence to the not uncommon and certainly respectable view that the old conceptual problems in QM are simply carried over into QFT, where they are obscured by the mathematically more sophisticated formalism. Allied to this view is the contention that QFT, in turn, raises no new foundational issues that are not already present in QM. If all of this is so, it none the less needs to be established in detail. In doing so, careful consideration must be given to the viewpoint according to which at least some of the problems in QM, e.g. the nature of wave–particle duality, receives ultimate clarification in QFT. It can also be argued that serious new difficulties arise in the foundations of QFT, e.g. in connection with renormalization procedures, the status of virtual particles, and the question of particle localizability, which have no real counterpart in the older theory.

Those physicists and philosophers who have contributed studies in the history and foundations of QFT would almost certainly be the first to agree that such issues have to date enjoyed far less attention than they deserve. Thus, the objective of this volume of essays is to delineate and examine a range of topics in the foundations of QFT that might be considered worthy of further study.

Part I is entitled 'Quantum Field Theory as an Object of Philosophical Study: Two Views'. It contains essays by Michael Redhead ('A Philosopher Looks at Quantum Field Theory'), and James Cushing ('Foundational Problems in and Methodological Lessons from Quantum Field Theory'). In a seminal 1982 paper, Redhead analysed in considerable detail the metaphysical implications of QFT. His present work is an improved version of that paper, in which besides presenting his own re-examination of several issues, he replies to objections raised by several commentators of the paper, notably Robert Weingard and Paul Teller. As in his earlier work, Redhead lists and attempts to answer eight central questions in the foundations of QFT. They concern the particle/field distinction, wave–particle duality, the nature of the vacuum, the role of quantum statistics, particle species unification, and other related issues. The upshot of the essay is effectively to underline the importance and novelty of metaphysical issues arising out of the theory, justifying Howard Stein's 1970 remark that QFT is, or should be, 'the contemporary locus of metaphysical research'.

Cushing, in his essay, is far less convinced that QFT indeed introduces any strong reasons for shifting this locus away from QM. In his view, neither wave–particle duality, nor even the conceptual problem of particle creation and annihilation, are fundamentally affected in the transition from QM to QFT. An examination of these and other issues leads Cushing to conclude that although QFT may be of considerable interest for those studying the dynamics of theory growth and scientific practice, it holds out little that is new and challenging from the point of view of foundational studies.

Part II is entitled 'The Problems of Virtual Particles and Renormalization'. 'Virtual' mechanisms in quantum theory really pre-date QFT. In Louis de Broglie's original 1923 matter-wave theory, the wave group, and not the individual superluminal phase wave, was considered real, but somehow the latter did—virtually—all the work in the theory. In 1924, the Bohr–Kramers–Slater theory of (explicitly) virtual radiation appeared, and when Erwin Schrödinger remarked that such terminology was merely playing with words, he was essentially raising the question of the reality of the BKS mechanism.

This particular issue died with the early demise of the theory, but in more recent times some philosophers of physics have been given to Schrödinger-like doubts about the virtual particles that play such an important role in the standard interpretation of perturbation methods in QFT.

In his essay ('Virtual Particles and the Interpretation of Quantum Field Theory'), Robert Weingard argues that if certain elements of the orthodox interpretation of states in QM are applicable to QFT, then it must be concluded that virtual particles cannot exist. This follows from the fact that the transition amplitudes correspond to superpositions in which virtual particle type and number are not sharp. Weingard argues further that analysis of the role of measurement in resolving the superposition strengthens this conclusion. He then demonstrates in detail how in the path integral formulation of field theory no creation and annihilation operators need appear, yet virtual particles are still present. This analysis shows that the question of the existence of virtual particles is really the question of how to interpret the propagators which appear in the perturbation expansion of vacuum expectation values (scattering amplitudes). Finally, Weingard examines the so-called Fayddeev–Popov ghost fields in gauge theory which violate the spin-statistics theorem. He argues that they are fictitious not because, like other virtual processes, they are associated with internal lines in Feynman diagrams, but because they can be transformed away in an appropriate sense.

In the following essay ('Parsing the Amplitudes'), Rom Harré comes to the defence of the now-battered virtual particle, in the course of providing a more general view of the nature of quantum (field theoretic) reality. Harré begins with the uses and perils of Arthur Miller's notion of visualizability in fundamental physics, and remarks in this context on the importance of the 'iconic' style of representation in Feynman diagrams in the QFT research programme. He argues, however, that in opting for a corpuscular language in the interpretation of internal states, physicists are also, and perhaps more directly, influenced by the exigencies of actual material practice in the laboratory. This leads Harré to a discussion of dispositional concepts, and more specifically of Gibsonian 'affordances', which he connects with the familiar notion of the Bohrian 'phenomenon'. To talk of virtual particles is then to talk of affordances, where the corpuscular aspect of the description follows from the nature of the track-like phenomena in high-energy physics experiments. The message here is that virtual particles are indeed different from their real counterparts (and in some cases historically precede them), but they are just as philosophically respectable.

Second- and higher-order corrections to solutions of perturbation methods in QFT are divergent, and the infinities have come to be successfully removed by the technique of renormalization. The word 'successfully' here refers to both the enormous accuracy of empirical predictions (to over eight decimal places in the cases of the Lamb shift and the electron magnetic moment) that arise out of the procedure, and the important connection between the renormalizability condition and the imposing of gauge symmetries. But can such infinities be discarded in any way that does not raise serious doubts about mathematical propriety, and does not their very existence in the first place point to severe weaknesses in the foundations of the theory? In the last essay of this section ('Three Problems of Renormalization'), Paul Teller addresses these issues. Teller attempts in a series of steps to undress the renormalization procedure of its usual, daunting technical clothes, and reveal its bare logic in a form accessible to the non-specialist. He tries first to show that from a purely mathematical point of view at least, doubts concerning the consistency of the method are unfounded. He further discerns and critically evaluates three distinct attitudes in the physics community towards the physical/philosophical significance of the procedure.

Part III is entitled 'Covariance Principles in QFT', and the first essay is by Gordon Fleming ('Hyperplane-dependent Quantized Fields and Lorentz Invariance'). In this work, Fleming advances a general formalism for a non-local field in relativistic QFT, and presents a specific example in $1+1$ dimensional space-time. The novel feature of the formalism is the non-local dependence of dynamical variables of the system on spacelike hypersurfaces, rather than individual space-time points, in the Minkowski continuum. Fleming examines, in his introduction to the theory, the reasons why such hyperplane dependence does not feature in the standard treatments of single quantum mechanical particles, nor in that of many-particle systems. In doing so, he claims that Lorentz covariance cannot be rigorously satisfied for particles with spatially local properties in the former case, and that in the latter, although hyperplane dependence is not strictly compulsory, it may prove to be advantageous. It may eventually furnish the basis of a finite fundamental theory, or failing that, shed light on non-perturbative methods in a fundamental renormalizable local field theory.

In the second essay ('Gauge Theory and the Geometrization of Fundamental Physics'), Tian-Yu Cao examines the interesting connection between non-gravitational gauge interactions in QFT and general relativity. After an introductory section dealing with the rise of gauge theory in QFT, Cao discusses the development of the fibre-bundle version of gauge fields, and compares the non-trivial mixing of space-time and internal space indices with the mixing of space and time indices in Minkowski geometry. He argues that the concept of gauge fields is at root geometrical, and carries his analysis over to the cases of modern Kaluza–Klein theory and superstring theory. Thus despite the apparent incompatibility between the programme of QFT, involving quantum fields with local coupling and propagation of field quanta, and the geometrical programme of relativity, Cao concludes that the essence of gauge interactions in QFT is as deeply geometrical as that of gravity in general relativity.

Part IV is called 'Mathematical Foundations of Quantum Field Theory'. Following Wigner's 1939 definition of an elementary particle in terms of a certain irreducible representation of the Poincaré group, it was possible to derive, rather than merely postulate, the relativistic free-field wave equations for all possible stable particles. In his essay in this section, Ray Streater ('Why Should Anyone Want to Axiomatize Quantum Field Theory?'), explains how the programme of axiomatic QFT arose with Wightman's attempt to do for fields with interaction essentially what Wigner had done for free fields. The author starts with a critique of the historical development of the Dirac equation, and then shows how by 1936, the existing relativistic free quantized fields adhered to certain proto-axiomatic desiderata, such as positive transition probabilities, no negative-energy solutions, primitive causality, locality, etc. The advent of theories of the nuclear force however raised new problems, giving rise to a host of models involving non-perturbative methods whose predictions differed significantly. Wightman's axiomatic approach, which relied on a number of the desirable features of the free field, was designed to separate the acceptable theories from the unacceptable ones. Streater discusses the ensuing progress made in this programme, the current status of the post-1972 triviality predictions, and the problem of fitting quantum chromodynamics into the axiomatic scheme.

The second essay is by Simon Saunders ('The Algebraic Approach to Quantum Field Theory'). The author examines the nature and historical roots of the abstract approach to QFT provided by the so-called C*-algebras. In such an approach, a distinction arises between abstract structures and their concrete representations, one which in Saunders's view requires further interpretation. He argues that no concrete representation captures the full scope of the theory, and is particularly interested in the significance of the availability of non-Fock representations. This leads to an exploration of the relevance of such representations in the high-energy regime and for the traditional problem of measurement in quantum theory. The paper also provides some insight into the far-reaching relationship between statistical mechanics and QFT.

The Sub-Faculty of Philosophy at the University of Oxford has in recent years been organizing an annual seminar in the philosophy of physics. Seven of the nine papers in this volume have their origin in contributions to the symposium on the philosophy of quantum field theory held in the Sub-Faculty from 30 May to 1 June 1986. We would like to thank Michael Redhead for his valuable advice in the course of organizing this symposium.

We also thank Angela Blackburn of Oxford University Press for her interest and help in overseeing the publication of this volume, and Simon Saunders for his help and particularly for his work in preparing the index.

Oxford, 1986

HARVEY BROWN
ROM HARRÉ

I
Quantum Field Theory as an Object of Philosophical Study: Two Views

1
A Philosopher Looks at Quantum Field Theory

MICHAEL REDHEAD

At the PSA Conference at Philadelphia in 1982 I gave a paper entitled 'Quantum Field Theory for Philosophers'. As a motto for the paper I took a quotation from Howard Stein: 'The quantum theory of fields is the contemporary locus of metaphysical research.'[1]

Let me begin by listing eight questions of a more or less metaphysical character which quantum field theory (QFT) might be thought to throw light on:

Q. 1. Can QFT be given a particle interpretation and indeed is there a formal underdetermination between field and particle approaches to the so-called elementary 'particles'?

Q. 2. Does QFT resolve the problem of wave–particle duality in quantum mechanics (QM)?

Q. 3. What is the nature of the vacuum in QFT?

Q. 4. What is the status of so-called virtual particles?

Q. 5. Does the theory of indistinguishable particles in QM necessitate a field treatment due to the way many-particle states are weighted in quantum statistical mechanics (QSM)?

Q. 6. Does QFT allow a distinction between matter and force?

Q. 7. In what sense has QFT achieved unification in the theory of elementary 'particles'?

Q. 8. Can the idea of creation and annihilation of particles be incorporated in classical mechanics as well as in QFT?

In my 1982 paper I attempted to answer these eight questions. Today I want to look at these questions again. In some cases I believe that what I said in 1982 was broadly on the right lines. In other cases I

---

[1] See Stein (1970), p. 285.

© Michael Redhead 1987.

have modified my position in the light of comments and criticisms or just by having thought harder about the questions!

In a moment I will tackle Q. 1. But as a preliminary that will also answer Q. 8, let us consider the case of classical point-particle mechanics. This is famously underdetermined as between a field and particle interpretation.[2] First some preliminaries:

A *particle theory* attributes to certain individuals (the particles) a variety of properties. These properties will include space-time location.

A *field theory* associates certain properties (the field amplitudes) with space-time points. Examples are the electromagnetic field, and Eulerian hydrodynamics as contrasted with Lagrangian hydrodynamics (a particle theory).

Already we have run into a number of metaphysical conundrums. What is an individual? For a particle individuation may be provided by some essential 'thisness' that transcends its properties. I will call this 'transcendental individuality' or TI for short. Or we may appeal to spatio-temporal (S-T) continuity of its trajectory. But this means we must be able to individuate space-time points. Do they possess TI? In field theories the space-time points play the role of the individuals. The problem of how they get individuated is then a very urgent one. Are they individuated by the fields which they carry? This view finds some support in general relativity, but this relies on an application of the identity of indiscernibles to space-time points! In QM, S-T individuation is not available. So if QM particles are to be treated as individuals then TI must be presumed. Any philosophical arguments against the admissibility of TI will then tell against a particle interpretation of QM.

Returning to the underdetermination thesis in classical mechanics: How does a single particle go from A to B?

Here are two answers: A material particle moves over from A to B carrying its individuality with it (on the TI assumption) as in Fig. 1.1;

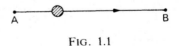

FIG. 1.1

---

[2] See the translation of Newton's 'De Gravitatione et Aequipondio Fluidorum' in Hall and Hall (1962), especially 138–40. Further discussion is given in Stein (1970).

A Philosopher Looks at Quantum Field Theory

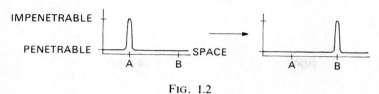

Fig. 1.2

or a property of impenetrability is assigned to a new spatial location B as opposed to A as in Fig. 1.2.

Can we break this up into a two-stage process? The spike at A (see Fig. 1.3) is annihilated leaving the vacuum configuration and then the spike at B is recreated out of the vacuum. But that is empirically wrong, since if we look we always find a particle between A and B. But if creation/annihilation followed each other sufficiently rapidly would we not get the *appearance* of continuous motion, just like the cinematograph screen?

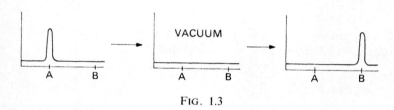

Fig. 1.3

The fact that we can never tell empirically which interpretation is right is a clear example of underdetermination by possible empirical data. This does not mean that field and particle ontologies are *heuristically* equivalent in generating theories that *are* testable.

Before giving our answer to Q.1. We are now in a position to interpolate a quick answer to Q.5. As we have seen, particles in QM cannot be individuated by S-T continuity of trajectory except in certain limiting cases of widely separated wave-packets. So we must invoke TI. But this leads to the following puzzle.

Consider the possible states available to two indistinguishable particles distributed among two distinct one-particle states denoted by a and b. Classical statistical mechanics (assuming TI) gives the possible ontologically distinct arrangements shown in Fig. 1.4. But in quantum statistical mechanics (iii) and (iv) are regarded as *one and the*

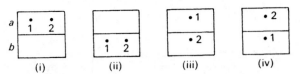

Fig. 1.4

same state for the purpose of assigning statistical weights. This is taken as showing that quantum particles do not possess TI, and must be treated like excitations of a field on which interpretation (iii) and (iv) *would* be the same state. But the argument is too quick. The statistical weights ascribed in QSM can be understood even if the particles do possess TI.

Consider the four possible product wave functions

$$\Psi_a(\mathbf{r}_1) \cdot \Psi_a(\mathbf{r}_2). \tag{1.1}$$

$$\Psi_b(\mathbf{r}_1) \cdot \Psi_b(\mathbf{r}_2). \tag{1.2}$$

$$\Psi_a(\mathbf{r}_1) \cdot \Psi_b(\mathbf{r}_2). \tag{1.3}$$

$$\Psi_b(\mathbf{r}_1) \cdot \Psi_a(\mathbf{r}_2). \tag{1.4}$$

These span a four-dimensional vector space which can equally well be spanned by the following wave functions:

Symmetric (S)
$$\Psi_a(\mathbf{r}_1) \cdot \Psi_a(\mathbf{r}_2). \tag{1.5}$$
$$\Psi_b(\mathbf{r}_1) \cdot \Psi_b(\mathbf{r}_2). \tag{1.6}$$
$$\frac{1}{\sqrt{2}}(\Psi_a(\mathbf{r}_1) \cdot \Psi_b(\mathbf{r}_2) + \Psi_b(\mathbf{r}_1) \cdot \Psi_a(\mathbf{r}_2). \tag{1.7}$$

Anti-symmetric (A)
$$\frac{1}{\sqrt{2}}(\Psi_a(\mathbf{r}_1) \cdot \Psi_b(\mathbf{r}_2) - \Psi_b(\mathbf{r}_1) \cdot \Psi_a(\mathbf{r}_2)) \tag{1.8}$$

Now for time-evolution under a symmetric Hamiltonian the symmetry character of the wave function is conserved.

So if we impose S or A as an initial condition then only one of the two states (1.7) and (1.8) is ever *available* to the system. This is why the statistical weights attaching to the *pair* of states get halved. In other words the statistical weights assigned in QSM can be regarded as arising from dynamical restrictions on the *accessibility* of certain states rather than on their ontological coalescence.

So at last we turn to Q. 1. First we consider the case of *non-interacting* fields. We distinguish *field quantization* from *second quantization*.

Field quantization

Here a 'real' classical field is regarded as a 'mechanical' system with an infinite number of degrees of freedom, which is then subjected to canonical quantization with the field amplitude interpreted as an operator. Consider the case of the real Klein–Gordon field. This satisfies the linear equation of motion

$$\left(\nabla^2 - \frac{1}{c^2}\frac{\partial^2}{\partial t^2} + \mu^2\right)\Psi = 0.$$

By Fourier analysis we can represent the motion of the field in terms of uncoupled normal modes, i.e. the problem of the motion of the field can be reduced to that of a system of independent harmonic oscillators. The quantization of the field is thus equivalent to quantizing an infinite collection of independent harmonic oscillators. But the solution of that problem is well known: The energy E given by

$$E = \sum_{\mathbf{k}} n_{\mathbf{k}}(\hbar\omega_{\mathbf{k}}) + \text{constant},$$

where

$$\text{constant} = \tfrac{1}{2}\sum_{\mathbf{k}}(\hbar\omega_{\mathbf{k}})$$

and $\omega_{\mathbf{k}} = c\sqrt{(\mu^2 + \mathbf{k}^2)}$ is the angular frequency of the mode. $n_{\mathbf{k}}$ are the integral eigenvalues of the operator $N_{\mathbf{k}} = a_{\mathbf{k}}^\dagger \cdot a_{\mathbf{k}}$ for the $\mathbf{k}$-mode, where $a_{\mathbf{k}}^\dagger$ and $a_{\mathbf{k}}$ are the usual raising and lowering operators for the harmonic oscillator with the properties

$$a_{\mathbf{k}}^\dagger |n_{\mathbf{k}}\rangle = \sqrt{(n_{\mathbf{k}}+1)} \cdot |n_{\mathbf{k}}+1\rangle$$
$$a_{\mathbf{k}} |n_{\mathbf{k}}\rangle = \sqrt{n_{\mathbf{k}}} \cdot |n_{\mathbf{k}}-1\rangle.$$

Neglecting the constant (the so-called zero-point energy) which we shall return to later, the spectrum for E can be interpreted as arising from $n_{\mathbf{k}}$ particles with energies $\hbar\omega_{\mathbf{k}}$.

But this identifies $\hbar\mathbf{k}$ with the momentum of the particle and $\hbar\mu/c$ with its rest-mass.

So we have here the simple equivalence: The excitation number $n_{\mathbf{k}}$ of the $\mathbf{k}$-mode is identified with the number of particles with momentum $\hbar\mathbf{k}$ associated with that state of the field.[3]

---

[3] Strictly the modes associated with particles of definite momentum correspond to travelling waves rather than standing waves. The treatment in the text glosses over this complication. For a more precise discussion see Redhead (1983).

Thus we get a particle interpretation of the quantized field. A particle or 'quantum' is just a quantized excitation of the field. But these 'particles' have no TI labels attached to them. They have no intrinsic individuality. They are a field version of *boson* particle theory. So starting with a field we end up with boson 'particles', but equally we can start with a theory of indistinguishable particles subject to symmetrized wave functions and subject it to a process of second quantization.

Second quantization

For non-interacting particles subject to an N-particle wave equation the energy spectrum is

$$E = \sum_i n_i E_i,$$

where E_i are the energies of the one-particle states.

Compare this formally with the energy spectrum of a collection of independent oscillators

$$E = \sum_i (n_i + \tfrac{1}{2}) E_i, \quad E_i = \hbar \omega_i.$$

But this is the result we would get by treating the one-particle wave equation *as if* it were a real field and subjecting it to field quantization. For example the Schrödinger field (which actually does not exhibit zero-point energy) is regarded as first quantization. We then treat it as though it were a classical field and field-quantize it and this mathematical trick is called *second quantization*.

But second quantization is more general than the N-particle wave equation because the constraint

$$\sum_i n_i = N$$

can be relaxed.

In second quantization we are dealing with Fock space, the direct sum of Hilbert spaces for no particles (the vacuum), one particle, two particles etc.

$$\mathscr{F} = \mathscr{H}_0 \oplus \mathscr{H}_1 \oplus \mathscr{H}_2 \oplus \ldots$$

For an operator Q on $\mathscr{H}_N$ which moves a particle from one one-particle state to another, we are 'factorizing' it into a product of creation and annihilation operators (see Fig. 1.5). Writing $Q = \xi \cdot \eta$ should be compared with the 'two-stage' version of the field

A Philosopher Looks at Quantum Field Theory 15

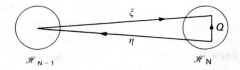

Fig. 1.5

interpretation of classical particle motion discussed above. Summarized schematically in Fig. 1.6.:

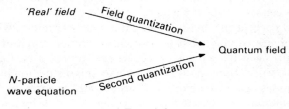

Fig. 1.6

But is the quantum field the same animal in the two cases? The question is raised even more sharply for the second quantization of fermions where anticommutator brackets replace commutator brackets in the quantization. The question is: Are some elementary 'entities' *really* particles and others *really* field excitations (quanta)?

Here is a summary of the main arguments in the literature on this point.

1. The Schrödinger field amplitude is not Hermitean and hence is not an observable, unlike, for example, the electromagnetic field. But the complex field can always be regarded as a pair of real fields, and in relativistic theories these are associated with the doubling up of particles and antiparticles carrying opposite charges.

For real fields individual Fourier components are not Hermitean, nor are the creation and annihilation operators. But in all cases Hermitean observables representing energy density, momentum density, charge density etc., *can* be constructed.

2. Observables must commute at spacelike separation for reasons of causality. So fermion fields, which anticommute, cannot be observables. True, but again we can construct observables such as

energy densities, which do commute at spacelike separation, and this is what the empirical content of the theory is about.

3. The fermion classical limit is a particle theory while the boson classical limit is a field theory. This may be true, but is not decisive away from the limit. In addition it is not clear that a bosonic particle classical limit cannot also be defined.

4. Massless fields do not define particles due to localization problems. This is related to the fact that massless quanta of low energies are very easy to create. But this does not seem decisive for not regarding the quanta (such as photons for example) as particles.

These are the main arguments for dividing the elementary entities into two classes, genuine particles and field quanta.

If we do accept the division along the boson–fermion divide, then with reference to Q. 6 we can distinguish fermionic matter (quarks and leptons) which are made up of particles from bosonic force fields (the gauge fields). Some doubt is thrown on this distinction by the fact that fermions can mediate interactions between bosons (see Fig. 1.7). So in some sense fermions are also connected with force (between field quanta!).

FIG. 1.7

Another remark often made in connection with Q. 6 is the role of supersymmetry in blurring the distinction between bosons and fermions. This now connects directly with Q. 7. It can be argued that internal symmetry transformations can only be given a meaninful *active* inerpretation, and if we regard as a sufficient and perhaps necessary condition for ontological unification the possibility of a *passive* interpretation, then (on the assumption of necessity) we can argue that internal symmetries do not signal ontological unification, but only a relatedness between the particles subsumed under the symmetry. In no sense does supersymmetry tell us that bosons *are* fermions and vice versa, ontologically speaking. We may remark here on the relevance of the geometrical Kaluza–Klein interpretation of

A Philosopher Looks at Quantum Field Theory

grand unified theories. If this is, in reality, the case, then grand unified multiplets could be regarded as ontologically unified.[4]

Returning to Q. 1, if we reject a distinction between genuine particles and field quanta as between one entity and another, have we got a situation of equivalence or underdetermination for all elementary entities? They can *eiher* be thought of as particles *or* as field quanta. This suggestion is directly relevant to Q. 2. We do not have to use particles or waves in complementary situations. Either can be used in all situations. This is sometimes claimed, following Dirac, to resolve the problem of wave–particle duality.[5]

In 1982 I argued for the underdetermination thesis, but denied that it solved the wave–particle duality problem. Now I am not so sure about the first point, but would still argue for the second, that is to say if we *did* have underdetermination, then one still would not have resolved wave–particle duality. The point is simply that in QFT we continue to have incompatible non-commuting operators to provide complementary pictures. In particular N_k does not commute with $\Psi(\mathbf{r})$, i.e. a representation in which N_k is diagonal (the particle picture) is complementary to one in which $\Psi(\mathbf{r})$ is diagonal (the wave picture). States with sharp $\Psi(\mathbf{r})$ can be expanded as superpositions of states with variable particle number, but that is no different from the usual transformation theory between representations in ordinary QM.

Let us revert now to the underdetermination thesis. To provide a particle interpretation of everything that happens in QFT we have to allow for *superpositions* of states with different particle numbers. In 1982 I defended this in terms of what I called an *extended* particle interpretation. But arguments of Weingard and Teller[6] have now persuaded me that extended particle interpretations are not really particle interpretations at all.

The problem is posed most sharply by turning to Q. 3. What is the nature of the QFT vacuum? In the vacuum state all the n_k are zero, so there are no particles, but there is still plenty going on, as evidenced by the zero-point energy already referred to. This arises from the quantum mechanics of the harmonic oscillator. Having N_k sharply zero is incompatible with the amplitude of the oscillation being sharply zero. The zero-point energy reflects vacuum fluctuations in the field amplitude. These produce observable effects such as the

---

[4] For further discussion see Weingard (1984) and Redhead and Steigerwald (1986).
[5] Cp. Dirac (1927), p. 245.
[6] Weingard (1982) and Teller (1983).

Lamb shift, the anomalous magnetic moment of the electron, the Casimir effect, etc. But to understand these effects we have to take as genuine physically significant observables, quantities like field amplitudes which are not diagonal in the particle interpretation. I am now inclined to say that vacuum fluctuation phenomena show that the particle picture is not adequate to QFT. QFT is best understood in terms of quantized excitations of a field and that is all there is to it. In 1982 I stressed the heuristic fertility of the field point of view. I would now be inclined to add an ontological ingredient here.

I now want to turn to the question of *interacting* fields in support of these remarks. A typical scattering process can be pictured as in Fig. 1.8. In the initial IN state we have freely-moving particles (or perhaps better quanta.) Then there is interaction and finally we are interested in the transition amplitude to a final OUT state again comprising freely-moving particles.

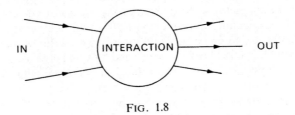

FIG. 1.8

This defines the S-matrix

$\langle \text{OUT}|S|\text{IN}\rangle$

Note here two important points:

1. We are working in the Schrödinger picture. In the more usual Heisenberg picture the $|\text{IN}\rangle$ and $|\text{OUT}\rangle$ states are defined as time-independent *collision states* and the S-matrix in the Heisenberg picture is defined by $\langle \text{OUT}|\text{IN}\rangle$.

2. When we said the particles in the IN and OUT states were freely moving this neglected their self-interactions. This leads to the problem of so-called wave-function renormalization in evaluating the contributions of Feynman diagrams to the transition amplitudes. This problem is best treated (compare the so-called LSZ approach)[7]

---

[7] For a comprehensive discussion see Bjorken and Drell (1965).

in the Heisenberg picture. We shall ignore this complication from now on. So $|IN\rangle$ and $|OUT\rangle$ are eigenstates of H_0, the free-particle Hamiltonian, while in the interaction region the states develop in time under the full Hamiltonian H. We shall suppose that the eigenstates of H_0 form a complete set for the Hilbert space on which H operates. Under this assumption (of asymptotic completeness)[8] we can expand the state at time t as

$$|\Psi(t)\rangle = \sum_n C_n(t)|\phi_n\rangle, \tag{1.1}$$

where $|\phi_n\rangle$ are the eigenstates of H_0.

If $|\Psi(-\infty)\rangle = |IN\rangle$, then the transition amplitude to the OUT state $|\phi_n\rangle$ is simply given by $C_n(\infty)$.

In general $C_n(\infty)$ is calculated by perturbation theory and yields a description in terms of a coherent sum of contributions from the relevant Feynman diagrams of all orders. That is just mathematics. The intermediate states or virtual particles identified with *internal* lines of the Feynman diagrams arise from expanding the state or propagator in the presence of interaction in terms of the state or propagator in the absence of interaction.

Thus in terms of propagators, and considering non-relativistic Schrödinger time-evolution, the basic Feynman integral equation in the presence of an interaction potential V is simply

$$K = K_0 + K_0 V K,$$

which can be rewritten

$$(1 - K_0 V) K = K_0$$

or

$$K = (1 - K_0 V)^{-1} \cdot K_0 = \sum_{n=0}^{\infty} (K_0 V)^n \cdot K_0.$$

K in this equation is the propagator in the presence of interaction and K_0 the propagator in the absence of interaction.
So

$$K = K_0 + K_0 \cdot V \cdot K_0 + K_0 \cdot V \cdot K_0 \cdot V \cdot K_0 + \ldots$$

or, diagrammatically, as in Fig. 1.9. The connection between K_0 and the $|\phi_n\rangle$ is

$$\langle 2|K_0|1\rangle_{\text{Df}} = K_0(2,1) = \sum_n \phi_n^*(\mathbf{r}_1)\phi_n(\mathbf{r}_2) e^{-i/\hbar \cdot E_n(t_2-t_1)} \cdot \theta(t_2-t_1),$$

---

[8] Strictly asymptotic completeness is often used in a different sense, the completeness of the $|IN\rangle$ and $|OUT\rangle$ states in the Heisenberg picture.

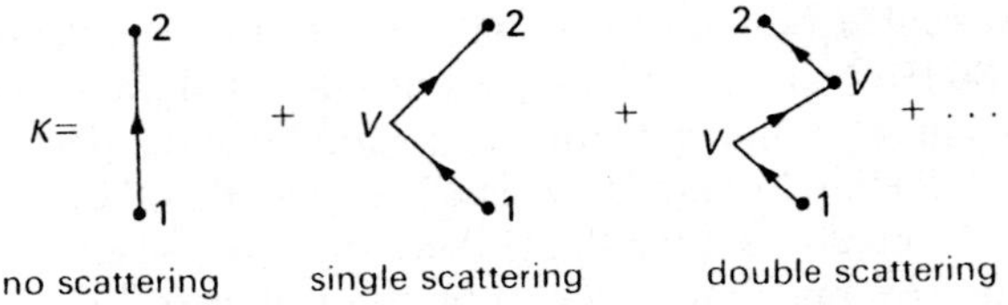

$$K = \;\underset{\text{no scattering}}{\big|_{1}^{2}} \;+\; \underset{\text{single scattering}}{V\diagup\!\!\!\diagdown_{1}^{2}} \;+\; \underset{\text{double scattering}}{V\diagup\!\!\!\diagdown V \diagdown_{1}^{2}} \;+\; \ldots$$

Fig. 1.9

where $\theta(t) = \begin{cases} 1, t > 0 \\ 0, t < 0 \end{cases}$ and E_n is the energy eigenvalue associated with $|\phi_n\rangle$.

In relativistic theories, such as QED, in terms of momentum space, energy and momentum is conserved at each Feynman vertex but the internal lines do not lie on the mass-shell. This is why they are called virtual as opposed to real particles.

Let us discuss Q. 4, in terms of the expansion (1.1) at times t in the interaction region, i.e. when the particles are interacting with one another. (1.1) is a mathematical expansion, rather like Fourier analysing the motion of a violin string. It can only be cashed out physically in terms of probability amplitudes for observing $|\phi_n\rangle$ at time t. But this measurement would require us to switch off the interaction at time t. But this is something that we cannot physically do. So (1.1) is just a mathematical expansion with no direct physical significance for the component states.[9] To invest them with physical significance is like asking whether the harmonics really exist on the violin string?

I have drawn an analogy with the motion of a violin string. But there is a very significant difference in the problem we are discussing. For small displacements the violin string is a linear problem, so we can introduce a normal mode analysis. But for the coupled fields the equations of motion are in general non-linear so unlike the free-field case we cannot reduce the problem to independent harmonic oscillators (the normal modes).

Hence we cannot introduce any analogue of the particle number operator for the coupled fields. To put this another way we *can*

---

[9] It may be objected that H_0, as an Hermitean operator, is measurable in the presence of interaction. This relies on the supposed 1:1 correspondence between Hermitean operators and measurable quantities in QM. The onus on someone defending such a view would be to specify how a measurement of H_0 could, operationally speaking, be carried out without switching off the interaction.

introduce particle numbers for the asymptotic states (free fields) but these operators do not commute with the full Hamiltonian, so the asymptotic particle number for the IN state is not conserved. In terms of the expansion (1.1), $|\Psi(-\infty)\rangle$ has a definite number of free-field particles, but $|\Psi(t)\rangle$ is in general a superposition of free particle number eigenstates, i.e. it only admits a particle interpretation if we allow what I called before the extended particle interpretation.

In summary, in scattering problems, the IN and OUT states are represented as eigenstates of a particle number operator. But $|\Psi(t)\rangle$ for any t other than $-\infty$ cannot be given a direct particle interpretation.

Note that this is also true for $|\Psi(+\infty)\rangle$ which must not be confused with $|OUT\rangle$, which is just one particular *component* of $|\Psi(+\infty)\rangle$. The reason why $|\Psi(+\infty)\rangle$ has nevertheless a *more* physical particle interpretation than the general $|\Psi(t)\rangle$ is that the interaction at $t = +\infty$ has been effectively switched off for us by the scattering.

So collision states are *related* to particle states in the sense that they develop out of $|IN\rangle$ which is a particle number eigenstate, a quantized excitation of the vacuum state $|0\rangle$. But QFT also allows states which are not apparently related to particle states even in this tenous way.

Firstly there is the question of (stable) bound states. These can only be analysed in terms of superposition. Unlike collision states their backward time-evolution does not lead to free particle states. But in general by bringing in *other* field interactions we can disintegrate bound states and generate asymptotically free particle states, so there is a *connection* at still further remove than is the case for collision states, with free particle states. But for bound states which exhibit confinement even this possibility is precluded. A candidate for this situation is the popular quark QCD model of the hadrons. The quarks can never be made *manifest* as particles.[10] In what sense then are they particles? (Confinement is made plausible by asymptotic freedom in momentum space. But this could be interpreted as saying that at very short distances of separation quarks can be regarded as free particles. But note that this freedom is strictly achieved only in the limit at which the two particles actually coincide; but is that really a situation where we can properly talk about two independent particles?)

---

[10] Cp. Redhead (1980) for further comments on this problem.

As a final example I mention the case of soliton solutions[11] of field equations which arise in gauge theories which exhibit spontaneous symmetry breaking and the associated degeneracy of the vacuum. In all our discussion of the free-field case we have introduced particles as quantized excitations of a unique vacuum state. But soliton states map non-trivially onto the vacuum manifold at spatial infinity. In a sense they ride not on a single vacuum but on a 'knot' in the vacuum manifold. In what sense solitons and other topological entities such as instantons, t'Hooft–Polyakov monopoles, and so on, can be regarded as particles in the same sense as free-field quanta remains an open question.

In my 1982 paper I finished by introducing a category of ephemerals to be contrasted with the familiar substantial continuants of classical physics. My critics claimed that ephemeral was just another name for a field or process phenomenon. If field is drawn widely enough to include quantum fields, with their particle grin, as I expressed it, then I agree. The point I was making was just that quantum field theory has more affinity with the Heraclitean Flux than with the Parmenidean One.

REFERENCES

BJORKEN, J. D., and DRELL, S. D. (1965): *Relativistic Quantum Fields* (New York: McGraw-Hill).

DIRAC, P. A. M. (1927): 'The Quantum Theory of the Emission and Absorption of Radiation', *Proceedings of the Royal Society of London* A**114**, 243–65.

HALL, A. R., and HALL, M. B. (1962): *Unpublished Scientific Papers of Isaac Newton* (London: CUP).

RAJARAMAN, R. (1975): 'Some Non-Perturbative Semi-Classical Methods in Quantum Field Theory (A Pedagogical Review)', *Physics Reports* **21C**, 227–313.

REDHEAD, M. L. G. (1980): 'Some Philosophical Aspects of Particle Physics', *Studies in History and Philosophy of Science* **11**, 279–304.

——(1983): 'Quantum Field Theory for Philosophers', in P. D. Asquith and T. Nickles (eds.), *Proceedings of the 1982 Biennial Meeting of the Philosophy of Science Association* **2**, 57–99.

—— and STEIGERWALD, J. S. (1986): 'Ontological Economy and Grand Unified Gauge Theories', *Philosophy of Science* **53**, 280–1

---

[11] See Rajaraman (1975) for a useful survey of soliton physics.

STEIN, H. (1970): 'On the Notion of Field in Newton, Maxwell and Beyond' in R. H. Stuewer (ed.), *Historical and Philosophical Perspectives of Science*, Minnesota Studies in the Philosophy of Science, 5 (Minneapolis, Minn.: University of Minnesota Press), 264–87.

TELLER, P. (1983): 'Comments on the Papers of Cushing and Redhead' in P. D. Asquith and T. Nickles (eds.), *Proceedings of the 1982 Biennial Meeting of the Philosophy of Science Association* 2, 100–11.

WEINGARD, R. (1982): 'Do Virtual Particles Exist?' in P. D. Asquith and T. Nickles (eds.), *Proceedings of the 1982 Biennial Meeting of the Philosophy of Science Association* 1, 235–42.

——(1984): 'Grand Unified Gauge Theories and the Number of Elementary Particles', *Philosophy of Science* 51, 150–5.

2
Foundational Problems in and Methodological Lessons from Quantum Field Theory

JAMES T. CUSHING

One can consider the developments in quantum field theory from the point of view of foundational problems posed about physical reality at both the epistemological and ontological levels, or as a 'laboratory' to study the actual methodology of the scientific enterprise. I shall address both of these issues in this paper. My contention is that at the foundational level there is essentially nothing new in quantum field theory that is not already present in non-relativistic quantum mechanics. This is quite different in emphasis from the view expressed by Redhead (1983a). Even in quantum mechanics most of the difficulties which arise are generated by the realist's inclination to take the formalism too literally as representing physical reality, often in a modified or distorted classical view. More serious consideration of Bohr's philosophical framework of complementarity may prove fruitful for coping with this problem, which is not new to quantum field theory, being already the central interpretative problem of quantum mechanics. On the other hand, as a source to illustrate how theories are constructed and established in actual scientific practice, quantum field theory does provide a wide range of useful examples. Here too, although there are many technically fascinating topics (such as renormalization), quantum field theory is pretty much (methodological) business as usual. Much more interesting, essentially new, methodological departures are seriously proposed (with varying degrees of credibility) by some rival or alternative developments in modern theoretical physics, such as the S-matrix programme, with its bootstrap conjecture, or cosmology, with various forms of the

---

© James T. Cushing 1987. The research reported in this paper was partially supported by NSF grant SES-8318884.

anthropic principle (Gale 1983) and with the inflationary universe scenario (Guth 1981).

Let me begin by considering some of the more obvious candidates for foundational problems raised by quantum field theory over and above those already posed by non-relativistic quantum mechanics. Teller (1985a) has recently argued that there are essentially no new problems and my comments here will largely agree with this conclusion of his, perhaps even downgrading a little the importance he attaches to his one candidate for a new problem—superposition of particle states. In quantum field theory we face the alternative, often opposing, paradigms of particles vs. fields (or waves). Although Dirac (1927) is often cited as having established the equivalence of these two interpretations (Redhead 1983a), the future of quantum field theory lay with the field as *the* primary entity, the particle aspect being pushed further into the background (Heisenberg and Pauli 1929; 1930; Weinberg 1977; 1985b. The particle concept survives as the quanta of the field or possibly as 'knots' in the field. It appears simplest and quite tempting to take the (operator) field (one for (all) electrons, another for protons, etc.) as existing everywhere (Dyson 1953) and to see the particles as the associated field quanta, the effects of which can be detected upon observation. However, this is not a problem conceptually different from the wave–particle duality already present in ordinary quantum mechanics. In both cases one has competing pictures or images, the particle being dominant in non-relativistic quantum mechanics and the field in quantum field theory, and the art or learned skill of the practitioner is the ability to know when to employ and how far to push these paradigms in given circumstances.

Another central feature of quantum field theory is particle creation and annihilation. While it is true that an elaborate and computationally highly successful operator formalism has been developed to exploit this aspect of the theory, this same conceptual problem is already present in the original Bohr model (1913) of the atom in which photons, i.e. electromagnetic field quanta, are created (emitted by the atom in a 'downward' transition of an atomic electron) and annihilated (absorbed by the atom in a reverse transition). The basic mechanism by which these creation and annihilation processes occur remains just as mysterious for these (massless) photons as it does, say, for massive mesons. Teller (1985a) has made what seems to me the very fruitful suggestion that we see quantum field theory as working with superpositions of various particle states (just as ordinary quantum mechanics does with various states of a given system) and that a particular (definite) particle state is 'reduced' out of this superposition upon measurement, very much as in the traditional quantum-mechanical formalism. This makes the basic similarities between these conceptual problems in the two theories quite apparent. Thus, *superposition*, with the attendant riddles of entanglement and reduction, remains *the* central and generic interpretative problem of quantum theory. Furthermore, it appears as though the superposition principle for states is not restricted to microsystems alone, but can be observed for macrosystems as well (Leggett 1984). The Schrödinger equation may be exact, not just an approximation to some non-linear equation (Gisin 1984). The creation–annihilation operator formalism gives us no insight into this process. In Fine's (1982) apt phrase, quantum theory is '. . . the blackest of black-box theories: a marvellous predictor but an incompetent explainer'. One attempt to cope with this puzzle within a realist approach to scientific theories is to speak of weighted potentialities, dispositions, or implicit intermediate states (Redhead 1983b; Teller 1985a). As I understand Teller's (1985a) 'harmonic oscillator interpretation' of quantum field theory, his attempt is a (realist) move in this direction. Even if we assume that these potentialities are causally determined, do we really understand any better how a given result is produced in a given observation? Are this and similar circumlocutions (e.g. Rohrlich 1986) to provide a realist interpretation of quantum theory anything more than a definitional move to paper over our ignorance with a new set of words? (For example quantum realism is that realism which is consistent with quantum mechanics! What have we gained? Where is the structure, or the positive paradigm?)

Is understanding of physical processes possible without being able to tell a causal story of the sequence of events from 'cause' to 'effect'? And, if not, then what does causality mean in a quantum theory? Although one often employs the term *causality* in quantum field theory, and even has a definite mathematical statement of the (micro)causality condition, i.e. the vanishing of the commutator of two field operators which are evaluated at spacelike separated points, there is no notion of what constitutes causality, but rather a prohibition against violating a first signal principle. That is, causality in this context actually functions as a condition of non-acausality (Cushing 1986b). Once again, though, we see that a problem for quantum field theory, in this case causality, is just a variation on a difficulty already present in ordinary quantum mechanics.

The basic task we face in attempting to produce (a feeling of) resolution of these fundamental quantum riddles is to separate mere psychological acclimation from genuine understanding. That is, how much of our sense of understanding is only what Planck (1949, pp. 33–4) expressed long ago: 'A new scientific truth does not triumph by convincing its opponents and making them see the light, but rather because its opponents eventually die, and a new generation grows up familiar with it.' This same concern has been raised by Costa de Bearegard (1983, pp. 515–16) in his discussion of issues associated with the Bell theorems.

Hard paradoxes are resolved only by producing a new and adequate paradigm, in Kuhn's words. In physics, this implies the production of a new mathematical recipe (e.g. Copernicus's heliocentrism, or Newton's inverse-square law) and tailoring an explanatory discourse exactly fitting the mathematics (e.g. Einstein's interpretation of the Lorentz–Poincaré formulas; or, still better, Minkowski's).

This sort of 'explanation' is usually felt (and often for a long time) as itself paradoxical. Newton's action at a distance, Einstein's 'reciprocal' interpretation of the Lorentz contraction, have very often been deemed 'hardly explanations at all'.

What occurs in the 'paradox and paradigm' peripateia (or, in Kuhn's words, in a 'scientific revolution') is a victory of formalism over modelism. In the EPR case we do have, since many years, the formalism. We are at home with it for performing calculations, but not yet for viewing our world, and our relation to it.

These are *very* difficult questions, indeed, but not ones peculiar to quantum field theory. They must already be faced in standard non-relativistic quantum mechanics. There would appear to be no advantage to considering these foundational problems of quantum theory within the context of quantum field theory, with all of its attendant technical complications, rather than in the context of standard non-relativistic quantum mechanics, where the mathematics is simpler and more familiar and the problems of interpretation more easily brought into focus. After all of the effort that has gone into attempting to make 'sense' out of quantum mechanics, it should be fairly clear that no bending or straightforward modification of a classical realism will be sufficient to produce a satisfactory resolution of these problems. Perhaps it is time that philosophers of science attempted to take seriously Bohr's philosophical ideas concerning complementarity. Only quite recently have some philosophers (Teller 1981; Folse 1985; Howard 1986) tried to present Bohr's philosophy as a coherent whole, rather than as a contradictory set of philosophically naïve views. Complementarity must not be seen as a *principle* (which Bohr never claimed it to be), but rather as a general viewpoint or framework within which to present a coherent and unambiguous description of phenomena associated with an *objective* physical reality, even though from this we may not be able to represent the physical reality behind these observations (cf. Folse 1985, pp. 202–6).

Since I have set this as *the* central problem of quantum theory, let me comment on it a little. Classical realism postulates an independent physical reality possessing properties corresponding to the state parameters used in a theoretical description of observed phenomena. Thus, classical mechanical parameters, such as the position and momentum, or the mass and charge of a particle, are taken as representations (or pictures) of the properties actually possessed by a corresponding real object. Furthermore, this object is seen as the cause of the observed phenomena which themselves warrant a given theoretical interpretation. Such independently existing objects having these properties are posited by classical realism and this conceptual framework is consistent with observations made on a macroscopic scale. Einstein was a proponent of classical realism. Fine (1984, p. 114) has argued that Einstein saw such realism as a *programme* which must finally be judged successful or unsuccessful on the basis of its long-run historical record. This is an example of *entheorizing* (in Fine's terminology) a concept within a larger theoretical framework which must be judged as a whole.

Bohr, with his general framework of complementarity, shifted the emphasis to the (unambiguous) *description* of phenomena produced by the interaction of an observed object with an observing system. For him, the error of classical realism was to infer from the phenomena of observation the ontological properties of an independently existing object. These phenomena are *not* themselves representations of the objects behind (and causing, via interactions) the phenomena observed. Bohr cut the bridge or direct connection between observation properties and properties possessed by the independently existing object. He effectively scuttled Newton's third Rule for Reasoning in Book III of the *Principia* (Newton 1934, p. 398) whereby universal properties observed in the macrorealm are carried over into the microrealm. According to complementarity, the phenomenal object described depends upon the cut made or the line drawn between the observing system and the observed object. Bohr was not an anti-realist or instrumentalist since he accepted the existence of an independent reality behind the observed phenomena. But, for him, a theory, e.g. quantum mechanics, can no longer give a description of this independent reality, but only predict relations among the observed phenomena. The theory is an abstraction whose components, e.g. the state vector, do not represent properties of independent objects (as opposed to the case in classical mechanics). Recently, in this same spirit, Peres (1984) has attempted to interpret the state vector as a procedure for preparing or testing physical systems. Bohr's programme made statements about epistemology, but was far less explicit about an ontology. His framework allowed for the *possibility* of an underlying ontology, although providing no details. It appears to me as though Folse (1985) basically argues that we 'entheorize' this assumed substratum and judge it in the context of the detailed successes of the complementarity programme. A currently popular mode of illustrating how clear Bohr could be on some points and how nebulous on others, such as ontology, is to use his own rabbi story (Folse 1985, 258; Petersen 1985, p. 299; Howard 1986, p. 1).

In an isolated village there was a small Jewish community. A famous rabbi once came to the neighboring city to speak and, as the people of the village were eager to learn what the great teacher would say, they sent a young man to listen. When he returned he said, 'The rabbi spoke three times. The first talk was brilliant—clear and simple. I understood every word. The second was even better—deep and subtle. I didn't understand much, but the rabbi understood all of it. The third was by far the finest—a great and unforgettable experience. I understood nothing, and the rabbi himself didn't understand much either.'

It remains unclear just what to do for a 'quantum ontology'. The world is surely 'unknowable' in the sense of classical realism, provided we accept the standard framework of quantum mechanics. What assurance do we have that we will be able to penetrate to that postulated level behind the phenomena? And, if not, is this really so different from the picture of reality which Planck (1949, pp. 107–8) presented us with?

... [science] must set and continually keep its sights on the objective reality which it seeks, and in this sense science can never dispense with *Reality* in the metaphysical sense of the term. But the real world of metaphysics is not the starting point, but the goal of all scientific endeavor, a beacon winking and showing the way from an inaccessibly remote distance.

However, I have digressed too far and these are topics best left for discussion on another occasion, both because I do not have solutions for them and because the subject of this symposium is quantum field theory, not quantum mechanics.

So let me return to candidates for foundational problems peculiar to quantum field theory. One of the most evident technical difficulties which is a hallmark of quantum field theory is renormalization. However, I claim it is just that—a technical problem, not a foundational one. There appears to be no compelling reason to consider the infinities associated with the renormalization programme of quantum field theory as representing anything corresponding to physical reality. It is every bit plausible that these infinities are an indication of the shortcoming of standard quantum field theories. In order to make a field theory believable as a calculational tool, it was a practical necessity to find an operationally consistent means of isolating the infinities in the theory (not in *nature*) to obtain finite corrections to predictions of observable quantities. If that had *not* been *doable*, then these theories would have been useless for making serious and extensive numerical predictions. However, that an *ad hoc*, even though covariant, prescription was found indicates neither that the theory has been rendered mathematically consistent nor that these (infinite) renormalization corrections have any counterparts in physical phenomena. Therefore I disagree with Teller's (1985b) suggestions that fundamental conceptual revisions or concessions, or even a new mathematics, should be considered to cope with renormalization. I believe that here again, as in his harmonic oscillator analogy, too much has been made of formalism or of a formal analogy which was subsequently taken seriously and literally as a representation of physical reality. None of this should be seen as denying the important role the criterion of renormalizability has played as a restrictive and guiding principle in the construction of empirically successful quantum field theories (Weinberg 1977). A similarly useful principle has been that of (local) gauge invariance. In fact, the ultimate and perhaps paradoxical achievement of renormalizability as a restrictive principle in selecting candidates for physically acceptable field theories could turn out to be that there may be only one, i.e., a unique, superstring theory which may contain *no* infinite quantities to be renormalized away and which, in a suitable 'low-energy' limit, reduces to the local gauge field theories being so actively studied today by theorists and supported by the work of experimentalists. If such a *finite* superstring theory survives its tests, then the problem of renormalization will cease to exist (much as the ether became a non-problem by simply disappearing from the framework of the theory). Quite obviously, then, I suggest that renormalization just be bracketed as any type of foundational problem.

The fact that quarks have been postulated as being *in principle* unobservable as free particles is sometimes cited as a new type of explanatory move made in current quantum field theories, especially since scientists suggest taking the existence of quarks just as seriously as that of electrons (or protons, pions, etc.). Let me make two observations about this. First, the basic line of argument remains hypothetico-deductive and these entities' existence is inferred from their effects. The sense in which experimentalists 'see' protons or electrons is only somewhat less tenuous than that in which they claim to 'see' quarks. Second, and more importantly, quarks may not remain as fundamental entities in a theory, should superstring theories prove adequate. Therefore worry about the ontological status of quarks would seem premature, at the minimum. One can find much more fundamentally radical and potentially interesting suggestions for revisions in the rules of or criteria for epistemological support in the anthropic principle (Gale 1983) in its several forms: WAP, SAP, PAP, FAP (Barrow and Tipler 1985); or even CRAP (Gardiner 1986). Here, too, some of the more seemingly bizarre versions of this principle (especially the participatory anthropic principle, or PAP) arise because of a literal, realistic interpretation of quantum mechanics. Another new departure is the bootstrap conjecture of the S-matrix programme (Cushing 1985a and b). But, again, these do not seem fair game for discussion under the rubric of this symposium.

Therefore, as far as foundational problems are concerned, I claim that there is little, if anything, essentially new in quantum field theory over and above those interpretative problems already familiar from non-relativistic quantum mechanics. And, for quantum theory generally, the central problem is superposition. Difficulties arise when we attempt to give an essentially traditional, realist interpretation to the formalism of quantum mechanics. It appears reasonable that such theories can make restricted epistemological claims but that they should not be burdened with an ontology of the microrealm. This focus on epistemology, without a detailed underlying ontology, is quite compatible with, and even indicated by, the bootstrap conjecture of the S-matrix programme (Cushing 1985a). In fact, one could argue that quantum field theory should not be taken seriously as a *fundamental* theory and, hence, one ought not to seek foundational problems there. Weinberg (1985a; 1985b) has suggested that quantum field theory by itself may have no content, but may simply be a convenient way of implementing the axioms of S-matrix theory. This is the essence of his first 'folk theorem' (Weinberg (1985b). His second 'folk theorem' we might term the harmonic oscillator theorem by analogy with the result from classical mechanics that the small oscillations of 'any' mechanical system about an equilibrium configuration are simple harmonic. His 'theorem' says that any (correct) theory studied at energies far enough below its natural energy scale can be described by a simple effective field theory. Just think of the (non-renormalizable) Fermi theory of weak interactions, today seen as a low-energy approximation to the unified electroweak gauge theory with its massive intermediate vector bosons. The really fundamental theory may be an S-matrix type of theory. This is not wholly implausible since unified gauge theories may emerge as a limiting case of superstring theories, which in turn may be derivable from an S-matrix formalism (Balázs 1986)—a formalism which takes particles as the fundamental paradigm. That is, it may simply be premature to take quantum field theory as sufficiently fundamental to warrant an extended discussion of foundational problems associated with it.

In a similar spirit, it is worth noting that some recent developments in theoretical high-energy physics indicate that space-time may not be a fundamental property in nature, but only an approximation arising in the limit of high complexity. If this should turn out to be so, then many of the foundational studies of space-time by philosophers would seem misplaced. That is, one must be careful where he looks to choose foundational problems. For a start, he had better make certain that the theory or concept is a fundamental one, and not just some approximation.

In either quantum formalism, whether it be quantum field theory or S-matrix theory, the superposition postulate is assumed for state vectors and this is sufficient to produce the entanglement riddle for these states. Therefore, as I emphasized above, in any quantum theory it is superposition which generates *the* basic interpretative problem.

Even though quantum field theory may offer little new in the way of foundational problems, it does provide many illustrations of the methodology of modern scientific practice (Schweber 1984) and is a marvellous proving (or disproving) ground for models of theory selection (Cushing 1986c). In recent high-energy physics one sees how a theorist's intuitions are often led by the mathematics of a formalism rather than by the physics as in a previous era (Cushing 1983). Accommodation is an essential part of the process of providing epistemological warrant in science (Cushing 1984). Let me give a few examples. In the compound nucleus model, a general language was first constructed to allow a description of the phenomena of interest. Then, after this model (or theory) had been put into a form which made it quite adequate empirically, some very fundamental or overarching principle was sought which would make that model seem required or obviously necessary. Such a foundation was provided by Wigner (1955) on the basis of causality. What this actually amounted to was a demand that a first signal principle not be violated. Wigner's beautiful argument did not address the problem of causality in quantum physics, but rather drew out the implications of the requirement that an interaction not be able to make a particle (or wave-packet) travel at an arbitrarily high speed. This is another instance of an attempt by scientists to make a theory appear to be the only one possible. As Weinberg (1977 pp. 33–4) has expressed it: 'After all, we do not want merely to describe the world as we find it, but to explain to the greatest possible extent why it has to be the way it is.' Perhaps this simply reflects the need we have to fit everything into a grand—it is hoped unique—scheme, something reminiscent of Freud's (1965 p. 158) *Weltanschung*, or world-view. Of course, this move is not peculiar to quantum field theory, but recurs throughout the history of science. Certainly, one of the more extreme cases of this is provided by conjectured uniqueness of the world, based on the bootstrap hypothesis of the S-matrix programme. There the fundamental principle is conservation of probability (Cushing 1985a). A similar need to *impose* order and simplicity (real or imagined) is also evident in the philosophy of science.

Furthermore, this scenario from the compound nucleus model is an instructive case of the meaning of a term—here *causality*—being redefined by scientific practice (Cushing 1986b). Shapere (1984) has argued that such accommodations are quite common in and essential to actual scientific practice, and Nersessian (1984) has very effectively used the historical record to reconstruct this process for the concept of the (electromagnetic) field, beginning with Faraday, and proceeding through Maxwell, Lorentz, and Einstein. Developments in recent theoretical physics continue the story of the evolution of the field concept. Terms in scientific discourse do not have static meanings, but are defined and redefined within the context of their evolving usage.

A telling example of selectively ignoring the weaknesses of a 'successful' theory is provided by the now-fashionable talk about Landau (non-perturbative) singularities and the failure of renormalization to cope with these. The basic papers on this problem were written in the mid-1950s (Landau, Abrikosov, and Khalatnikov 1954; Landau 1955). The claim is that (even renormalized) perturbation theory must break down, i.e. become unreliable, at high enough energy. There is no resolution of this problem (assuming it actually exists) *within* QED. However, my own recollection is that this problem was not stressed until fairly recent times, *after* larger theories offered a resolution to it. For example gauge field theories, which are asymptotically free, have a running coupling constant which *decreases* with energy rather than increasing (as in quantum electrodynamics). As so often happens, embarrassments are shunted aside until they become victories for a successor theory. Similarly, the Fermi theory of weak interactions is an example of a (non-renormalizable) theory whose ('finite') predictions were known to break down at high energy. Yet this theory is now seen as a low-energy approximation to a 'better' theory (the standard electroweak gauge theory). This is in some ways similar to the present hope that quantum chromodynamics (QCD) will emerge as a low-energy limit to superstring theory. A possibility is that superstring theories, *if* they really do turn out to be finite, may be so complicated as to be of little use for actual calculations. Still, if they did provide a finite theory from which to obtain, say, QCD as a limiting case (with which we *can* calculate), the superstring theories would have served an important function. And, as I have mentioned above, superstring theory may also follow from (the rival) topological *S*-matrix theory programme (Balázs 1986).

Finally, a new dimension—the influence of social factors upon the scientific enterprise—has been stressed in recent years, and illustrated within the context of quantum field theory (Pickering 1984; Schweber 1984; 1985). Even if one has serious reservations about the 'strong programme' of the Edinburgh school (Bloor 1976), one must admit that the sociologists of knowledge have made implausible any story which would present science as following some necessary or unique course. It is not unlikely that the picture of the world science presents to us could be *very* different from what it happens to be today. This can be seen from the role that concepts from a 'dead-end' theory (by consensus of the scientific community)—the S-matrix programme—have had in providing the basis for a now actively pursued theory (superstrings) (Cushing 1985b; 1986a). Or, in a slightly more fanciful vein, let me pose a rhetorical question. If the Bohm–Vigier version of quantum mechanics had been pursued and developed before the Heisenberg–Schrödinger one and if Bell's theorem and its implications had been seen as soon as the standard quantum mechanics had been proposed, would standard quantum mechanics still be the theory of choice today or might we rather favour a deterministic theory with its independent classical reality? In the latter case, of course, there is no interpretative problem, but there is a foundational conflict with relativity (Bell 1986).

So, to conclude, I find the quantum field theory enterprise of most interest for studying the (ongoing) dynamics of scientific practice and knowledge construction, rather than as a source for new foundational problems. (For detailed examples see Cushing 1982; 1983; 1984; 1985a and b; 1986a,b, and c.) And, to the extent that one chooses to take quantum mechanics seriously, and as embodying the essential riddle of any quantum theory, one should seek a radical departure from the notions of classical realism. Patches and definitional moves have not proven sufficient to cope with these interpretative problems. Something much more fundamentally revisionary, such as Bohr's framework of complementarity, may be required. At the other extreme, though, the de Broglie–Bohm version of quantum mechanics can be extended to a perfectly workable quantum field theory (Bell 1986). One then has classical realism and determinism. The price, which may be small to non-existent in the practical sphere, is a violation of the usual condition of microcausality, but in such a way that superluminal communication remains impossible.

REFERENCES

Balazs, L. A. P. (1986): 'Could there be a Planck-Scale Unitary Bootstrap Underlying the Superstring?', *Physical Review Letters* **56**, 1759–62.

Barrow, J. D., and Tipler, F. J. (1986): *The Anthropic Cosmological Principle* (Oxford: OUP).

Bell, J. S. (1986): 'Introductory Remarks' and 'Quantum Field Theory without Observers' in E. R. Caianello (ed.), *New Avenues in Quantum Theory and General Relativity, Physics Reports* **137**(1), 7–9, 49–54.

Bloor, D. (1976): *Knowledge and Social Imagery* (London: Routledge & Kegan Paul).

Bohr, N. (1913): 'On the Constitution of Atoms and Molecules', *Philosophical Magazine* **26**, 1–25.

Costa De Beauregard, O. (1983): 'Running backwards the Mermin Device: Causality in EPR Correlations', *American Journal of Physics* **51**, 513–16.

Cushing, J. T. (1982): 'Models and Methodologies in Current Theoretical High-Energy Physics', *Synthese* **50**, 5–101.

—— (1983): 'Models, Theoretical High-Energy Physics and Realism' in P. D. Asquith and T. Nickles (eds.), *Proceedincs of the 1982 Biennial Philosophy of Science Association Meeting* **2**, 31–56.

—— (1984): 'The Content and Convergence of Scientific Opinion' in P. D. Asquith and P. Kitcher (eds.), *Proceedings of the 1984 Biennial Meeting of the Philosophy of Science Association* **1**, 211–23.

—— (1985a): 'Is there just One Posssible World? Contingency vs. the Bootstrap', *Studies in History and Philosophy of Science* **16**, 31–48.

—— (1985b): 'The S-Matrix Program—Anatomy of a Scientific Theory', University of Notre Dame preprint, Dec. 236 pp.

—— (1986a): 'The Importance of Heisenberg's S-Matrix Program for the Theoretical High-Energy Physics of the 1950s', *Centaurus* **29**, 110–149.

—— (1986b):'Causality as an Overarching Principle in Physics', in A. Fine and P. Machamer (eds.), *Proceedings of the 1986 Biennial Meeting of the Philosophy of Science Association* **1**, 3–11.

—— (1986c): 'The Justification and Selection of Scientific Theories', paper delivered at the Conference on Testing Theories of Scientific Change, Blacksburg, Va, 20–2 Oct.

Dirac, P. A. M. (1927): 'The Quantum Theory of the Emission and Absorption of Radiation', *Proceedings of the Royal Society of London* **A114**, 243–65.

Dyson, F. J. (1953): 'Field Theory', *Scientific American*, Apr., 57–64.

Fine, A. (1982): 'Antinomies of Entanglement: The Puzzling Case of the Tangled Statistics', *Journal of Philosophy* **79**, 733–48.

—— (1984): 'Einstein's Realism' in J. T. Cushing *et al.* (eds.), *Science and Reality* (Notre Dame, Ind: University of Notre Dame Press), 106–33.

FOLSE, H. J. (1985): *The Philosophy of Niels Bohr* (Amsterdam: North-Holland).
FREUD, S. (1965): *New Introductory Lectures on Psychoanalysis* (New York: W. W. Norton & Co.).
GALE, G. (1983): 'An Attempt to Explain Two Contemporary Divergences from Mainstream Philosophy of Physics' in P. Weingartner (ed.), *Abstracts of the 7th International Congress of Logic, Methodology and Philosophy of Science* (Salzburg), Vol. 4, 63–5.
GARDINER, M. (1986): 'WAP, SAP, PAP, & FAP', *New York Review of Books*, 8 May, 22–5.
GISIN, N. (1984): 'Quantum Measurements and Stochastic Processes', *Physical Review Letters* **52**, 1657–60.
GUTH, A. (1981): 'Inflationary Universe: A Possible Solution to the Horizon and Flatness Problems', *Physical Review* **D23**, 347–56.
HEISENBERG, W., and PAULI, W. (1929): 'Zur Quantendynamik der Wellenfelder', *Zeitschrift für Physik* **56**, 1–61.
—— and PAULI, W. (1930): 'Zur Quantentheorie der Wellenfelder, II', *Zeitschrift für Physik* **59**, 168–90.
HOWARD, D. (1986): 'What Makes a Classical Concept Classical? Toward a Reconstruction of Niels Bohr's Philosophy of Physics', paper delivered at the Bohr Centenary Symposium of the Boston Colloquium for the Philosophy of Science, Boston, 11 Mar.
LANDAU, L. D. (1955): 'On the Quantum Theory of Fields' in W. Pauli (ed.), *Niels Bohr and the Development of Physics* (London: Pergamon Press).
—— ABRIKOSOV, A. A., and KHALATNIKOV, I. M. (1954): 'The Removal of Infinities in Quantum Electrodynamics', *Doklady Akademii Nauk SSSR* **95**, 497–9, 773–6, 1117–20; **96**, 261–3.
LEGGETT, A. J. (1984): 'The Superposition Principle in Macroscopic Systems' in S. Kamefuchi *et al.* (ed.), *Proceedings of the International Symposium on the Foundations of Quantum Mechanics in the Light of New Technology* (Tokyo: Physical Society of Japan), 74–82.
NERSESSIAN, N. J. (1984): *Faraday to Einstein: Constructing Meaning in Scientific Theories* (Dordrecht: Martinus Nijhoff).
NEWTON, I. (1934): *Mathematical Principles of Natural Philosophy*, F. Cajori (ed.), 3rd edn. (Berkeley: University of California Press).
PERES, A. (1984): 'What Is a State Vector?' *American Journal of Physics* **52**, 644–50.
PETERSEN, A. (1985): 'The Philosophy of Niels Bohr' in A. P. French and P. J. Kennedy (eds.), *Niels Bohr: A Centenary Volume* (Cambridge, Mass.: Harvard University Press).
PICKERING, A. (1984): *Constructing Quarks: A Sociological History of Particle Physics* (Chicago: University of Chicago Press).

PLANCK, M. (1949): *Scientific Autobiography and Other Papers* (Westport, Conn.: Greenwood Press).
REDHEAD, M. L. G. (1983a): 'Quantum Field Theory for Philosophers' in P. D. Asquith and T. Nickles (eds.), *Proceedings of the 1982 Biennial Meeting of the Philosophy of Science Association* **2**, 57–99.
—— (1983b): 'Nonlocality and Peaceful Coexistence' in R. Swinburne (ed.), *Space, Time and Causality* (Dordrecht: D. Reidel), 151–89.
ROHRLICH, F. (1986): 'Reality and Quantum Mechanics', in D. M. Greenbergen (ed.), *New Techniques and Ideas in Quantum Measurement Theory*, Annals of the New York Academy of Sciences **480**, 373–381.
SCHWEBER, S. S. (1984): 'Some Chapters for a History of Quantum Field Theory: 1938–1952', in B. S. DeWitt and R. Stora (eds.), *Relativity, Groups and Topology III* (Amsterdam: North-Holland), 37–220.
—— (1985): 'Particle Theory in the 50's, an Historical Assessment', to appear in *Proceedings of the International Symposium on Particle Physics in the 1950s: Pions to Quarks*, Fermi National Accelerator Laboratory, 1–4 May.
SHAPERE, D. (1984): *Reason and the Search for Knowledge* (Dordrecht: D. Reidel).
TELLER, P. (1981): 'The Projection Postulate and Bohr's Interpretation of Quantum Mechanics' in P. D. Asquith and R. N. Giere (eds.), *Proceedings of the 1980 Biennial Meeting of the Philosophy of Science Association* **2**, 201–23.
—— (1985a): 'Prolegomenon to a Proper Interpretation of Quantum Field Theory', University of Illinois at Chicago Circle preprint, Sept.
—— (1985b): 'Three Problems of Renormalization', University of Illinois at Chicago Circle preprint, July.
WEINBERG, S. (1977): 'The Search for Unity: Notes for a History of Quantum Field Theory', *Daedalus* **106**(4), 17–35.
—— (1985a): 'Particles, Fields, and Now Strings', paper delivered at the Niels Bohr Centenary Symposium in Copenhagen, 7 Oct.
—— (1985b): 'The Ultimate Structure of Matter' in C. DeTar, J. Finkelstein and C.-I. Tan (eds.), *A Passion for Physics* (Philadelphia, Pa.: Taylor & Francis).
WIGNER, E. (1955): 'On the Development of the Compound Nucleus Model', *American Journal of Physics* **23**, 371–80.

II
The Problems of Virtual Particles and Renormalization

3
Virtual Particles and the Interpretation of Quantum Field Theory

ROBERT WEINGARD

In 'Do Virtual Particles Exist',[1] I argued that virtual particles do not, in fact, exist. I review that argument in section I of this paper. Then in section II I present the argument from a different point of view, emphasizing the role of the propagator, and in section III I look at the connection between the Fayddeev–Popov ghost fields and the problem of virtual particles.

I

In 'Do Virtual Particles Exist', I claimed that the reason it is believed that virtual particles exist is because of the appearance of annihilation and creation operators in the perturbation expansion of the S-matrix. Let us briefly review this. Writing our transition amplitude as $\langle f|S|i \rangle$, and $S = \Sigma_k S_k$, we have

$$\langle f|S|i \rangle = \sum_k \langle f|S_k|i \rangle.$$

In the case of electrodynamics, for example, the interaction Lagrangian, $\mathcal{L}_{1,x_\ell} \sim (\bar{\psi}\gamma^u A_u \Psi)_{x_\ell}$, $S_k \sim \Pi^k_{\ell=i} \mathcal{L}_{1,x_\ell}$ (suppressing the space-time integrations) and

Ψ contains a_k, which annihilates electrons of momentum k.
                   b_k^+, which creates positrons of momentum k.
$\bar{\Psi}$ contains a_k^+, which creates electrons of momentum k.
                   b_k, which annihilates positrons of momentum k.
A_u contains C_q^+, which creates photons of momentum q.
                   C_q, which annihilates photons of momentum q.

[1] In *Proceedings of the Philosophy of Science Association* (1982), ed. Peter Asquith and Thomas Nichols, (East Lansing, Mich.: Philosophy of Science Association), pp. 235–42.

© Robert Weingard 1987.

Let us look at the second-order term $S_2 \sim (\bar{\Psi}\gamma^u A_u \Psi)_x (\bar{\Psi}\gamma^u A_u \Psi)_{x'}$. In terms of the annihilation and creation operators, we get all the combinations resulting from

$$[(a_k^+ + b_k)(C_q + C_q^+)(a_{k'} + b_{k'}^+)]_x [a_{k''}^+ + b_{k''})(C_{q'} + C_{q'}^+)$$
$$(a_{k'''} + b_{k'''}^+)]_{x'}.$$

Suppose we are interested in electron–electron scattering. Our initial two-electron state is $|k_1 k_2\rangle$, and our final state $\langle k_3 k_4|$, so $\langle k_3 k_4 | S_2 | k_1 k_2 \rangle$ gives the second-order contribution to the transition amplitude to go from $|k_1 k_2\rangle$ to $|k_3 k_4\rangle$. Now only terms in S_2 that transform $|k_1 k_2\rangle$ into a state $\sim |k_3 k_4\rangle$ will contribute. For example, ignoring ordering problems,

$$[a_{k_2} a_{k_4}^+ C_q](x) [a_{k_1} a_{k_3}^+ C_q^+](x') = \hat{A}$$

will contribute because $a_{k_2} a_{k_1}$ annihilate the initial electrons, $C_q C_q^+$ create and destroy an intermediate (virtual) photon, and $a_{k_4}^+ a_{k_3}^+$ create the final state electrons. Thus

$$\hat{A} |k_1 k_2 \rangle \sim |k_3 k_4\rangle$$

so $\langle k_3 k_4 | \hat{A} | k_1 k_2 \rangle \neq 0$. Diagrammatically we can write this matrix element as shown in Fig. 3.1. Here we see clearly how the virtual photon arises from the action of the annihilation and creation operators C_q, C_q^+, in creating and annhilating particles (states) that do not appear in either the initial nor the final state. Now we know that a photon in an initial state can be annhilated by C_q, and one can

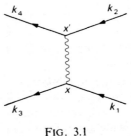

FIG. 3.1

FIG. 3.2

be created in a final state by C_q^+—this happens, for example, in compton scattering which to second order is shown in Fig. 3.2.

So the operators C_q, C_q^+ that here are destroying and creating physical (= on mass shell) photons in the initial and final states, are the *same operators* that, in our case of electron–electron scattering, act to create and destroy internal states. It seems natural, then, to interpret them as also creating and destroying 'real' photons in this case too, even though the photons never appear in the external states.

In spite of this, I argued in that earlier paper that virtual particles do not exist—or that virtual processes do not occur—for the following reason. In ordinary quantum mechanics, if the state ϕ of a system is $\phi = \Sigma_i a_i \chi_i$, where the χ_i are the eigenvectors of the property (observable) P, then according to the standard interpretation, P does not have a value in the system. Only if the system is part of an ensemble which, due to ignorance, we describe by the mixture $\rho = \Sigma |a_i|^2 \chi_i \rangle \langle \chi_i|$ would it be possible for P to be sharp (even this does not follow if P has degenerate eigenvalues). Going back to electron–electron scattering, besides the relevant terms in S_2, there are relevant terms in an infinite number of other S_k that contribute to the total amplitude $\langle k_3 k_4 | S | k_1 k_2 \rangle$, shown diagrammatically in Fig. 3.3. Since the amplitude is a sum of terms, the transition probability

$$|\langle f|S|i\rangle|^2 = |\sum_k \langle f|S_k|i\rangle|^2$$

is the square of a sum, which is characteristic of a superposition rather than the sum of squares that is characteristic of a mixture. Relying on the standard interpretation I concluded that neither the number nor kinds of virtual particles are sharp in the superposition represented by $\langle f|S|i\rangle$.

That is, it is not just that our amplitude $\langle f|S|i\rangle$ is a superposition. If the only diagrams that contributed to electron–electron scattering were those shown in Fig. 3.4 then although we have a superposition,

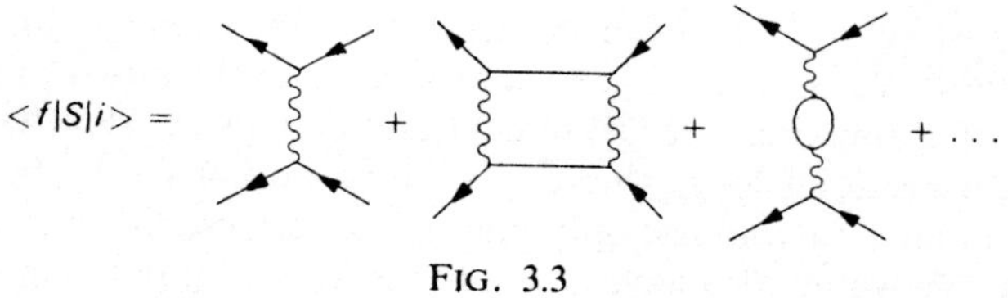

FIG. 3.3

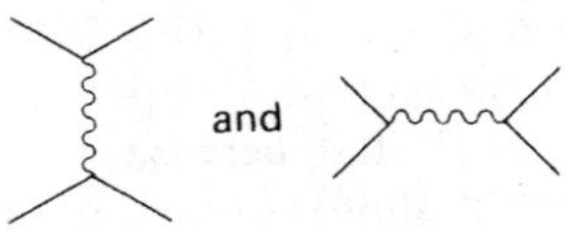

FIG. 3.4

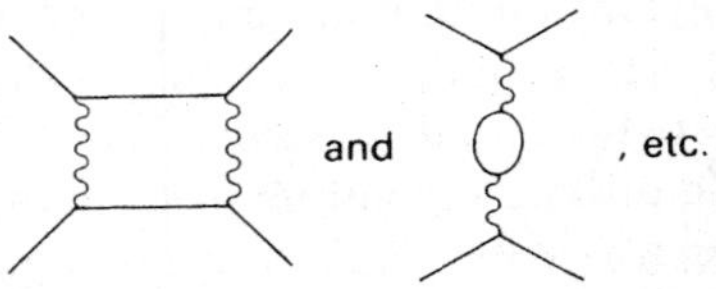

FIG. 3.5

the virtual particle number and type would be sharp. But this is not the case. The diagrams in Fig. 3.5 also contribute, so virtual particle type and number are not sharp. How then can we talk about there being virtual particles or processes which mediate the interaction?

Indeed, there is a related worry I did not mention in that earlier paper. Connected to the view about superposition mentioned above is a view about measurement—the familiar collapse of the wavepacket. In the context of our present discussion it boils down to this: the result of the interaction of the two electrons is the massive superposition $S|i\rangle$. As we have already mentioned, this contains not only the terms that contribute to $\langle f|S|i\rangle$ but also the terms relevant to all other possible final states. Let us indicate this symbolically by writing

$$S|i\rangle = \sum_\alpha |I_\alpha\rangle,$$

where α is a set of indices distinguishing all possible final states. Thus $\langle f|S|i\rangle = \langle f|I_{(f)}\rangle$, where $|I_{(f)}\rangle$ is the part of $S|i\rangle$ that contributes to $\langle f|S|i\rangle$.

So we have this massive superposition that results from the interaction. We only get a particular final state (here two outgoing electrons) by making a measurement and collapsing $S|i\rangle$. To hold that the transition $|i\rangle \to |f\rangle$ takes place by means of a virtual process we would have to hold not only that $|I_\alpha\rangle$ describes a definite virtual process. We would also have to hold that while before the measurement is made no definite virtual process has taken place, the measurement brings it about that $|i\rangle \to |f\rangle$ takes place by means of a particular virtual process. And this, even though the measurement occurs after the interaction takes place!

It might be objected that we can observe virtual particles, namely if we make a measurement while the interaction is taking place we will find some of the particles indicated by the Feynman diagrams. Well, first of all, this cannot be strictly correct because virtual particles are off mass shell while any observed particle will be on mass shell, but perhaps the measurement forces them onto the mass shell. Second, it does not follow that the particles we detect when we interrupt an interaction would have been there if we had not made the measurement, or were there just before the measurement. This is exactly analogous to our case of $\phi = \sum_i a_i \chi_i$. Before the measurement, P is not sharp. If I make a measurement of P and find p_i, that does not mean the value of P was p_i before the measurement because ϕ is a superposition of P eigenstates. That is the whole point: the measurement has changed the state of the system.

I have mentioned measurement for the sake of completeness—it is a natural development of the argument based on superposition. But now, having reviewed our earlier reasoning, I want to turn to the path integral formulation of quantum field theory.

II

Our original reason for thinking there might be virtual particles and processes in electrodynamic interactions came from our interpretation of the annhilation and creation operators that appear in $\mathscr{L}_1 \sim \bar{\Psi}\gamma^\mu A_\mu \Psi$. However, using path integrals, we can formulate field theory, and derive the Feynman rules (the Feynman diagram expansion) without introducing annhilation and creation operators. In this section I want to review how this works and what happens to virtual particles when we do this.

In the path integral formulation of field theory we generalize the result from ordinary quantum mechanics that

$$\langle q_f, t_f | q_i, t_i \rangle = \int \mathscr{D}q(t) e^{i\int_{t_i}^{t_f} L dt}, \quad L = L(q(t), \dot{q}(t)),$$

where $\langle q_f, t_f | q_i, t_i \rangle$ is the amplitude for a particle at q_i at t_i to be found at q_f at t_f, and the path integral is the 'sum' of the weighted action e^{iS}

over all paths between q_i, t_i and q_f, t_f. There are two key ideas needed which we can illustrate with ordinary quantum mechanics. One is that in the limit $t_i \to -\infty$, $t_f \to +\infty$, our path integral is proportional to the amplitude for a transition from the ground state at $t = -\infty$ to the ground state at $t = +\infty$,

$$\langle E_0, +\infty | E_0, -\infty \rangle = N \int \mathscr{D}q(t) e^{i\int_{-\infty}^{\infty} L\,dt}, \quad N = \text{constant}.$$

The other idea is that we get vacuum expectations of time-ordered products like $q(t_1)q(t_2)$ by adding a source term and differentiating. For example

$$\langle E+\infty | Tq(t_1)q(t_2) | E_0 -\infty \rangle \propto \frac{\delta}{i\delta j(t_2)} \frac{\delta}{i\delta j(t_1)}$$

$$\left\{ \int \mathscr{D}q e^{i\int_{-\infty}^{\infty} dt\{L+j(t)q(t)\}} \right\}_{j=0}$$

In passing over to the path integral formulation of field theory, I must emphasize that we will not be giving a perturbation expansion of the S-matrix (or the S-matrix) amplitudes $\langle f|S|i\rangle$—at least not directly. Instead we will compute the vacuum expectation values of the time-ordered products of the fields. Once we have these, it is easy to calculate the transition amplitude. For suppose we have a self-interacting scalar field and we are interested in calculating a scattering amplitude which we can picture (see Fig. 3.6). Thinking in terms of Feynman diagrams, our diagram (up to whatever order we calculate) has two parts. All the internal lines and vertices, and the external legs. Here we have four external legs which are on mass shell scalar particles, while the internal lines are the off mass shell 'virtual' particles. The vacuum expectation

$$\langle 0 | T(\phi_1 \phi_2 \phi_3 \phi_4) | 0 \rangle$$

expanded up to the relevant order gives us Fig. 3.7, where the four external legs represent free particle propagators. To get the scattering

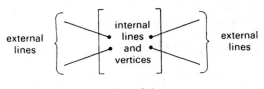

Fig. 3.6

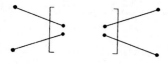

FIG. 3.7

amplitude (to that order) we have to remove the external propagators and replace them with external wave functions and some numerical factors. So it is the vacuum expectations of the time-ordered products that are important.

What we want to see, than, is how to get the perturbation expansion of the vacuum expectation value (VEV) without the use of annhilation and creation operators. It works like this, where we illustrate with the self-interacting scalar field. The Lagrangian density has the form $\mathscr{L} = \mathscr{L}_0 + V(\phi)$. $\mathscr{L}_0$ is the free field Lagrangian and $V(\phi)$ is the interaction term, which for our purposes we can take to be of the form $\lambda\phi^n$. Proceeding analogously to the case of quantum mechanics, our starting-point is

$$\langle 0+|0-\rangle_I^j = N \int \mathscr{D}\phi(x) e^{i\int d^4x(\mathscr{L}+j(x)\phi(x))}. \qquad (3.1)$$

This is the amplitude for the field to go from the vacuum at $t = -\infty$ to the vacuum at $t = +\infty$, under the influence of the interaction term and the source term. This gives us $\langle 0|T(\phi \ldots \phi)|0\rangle$ by differentiation with respect to $J(x)$ and then setting $J=0$.

The perturbation expansion then developes as follows. First we solve $\langle 0|0\rangle_{\text{free}}^j$,

$$\langle 0|0\rangle_f^j = N \int \mathscr{D}\phi \, e^{i\int d^4x(\mathscr{L}_0 + j(x)\phi(x))} \qquad (3.2)$$

$$= \langle 0|0\rangle_f^0 \, e^{-i/2 \int d^4x_1 d^4x_2 j(x_1)\Delta_F(x_1,x_2)j(x_2)}, \qquad (3.3)$$

where Δ_F is the Feynman propagator. Second, we include the interaction terms by noting that we can write the interaction term in (3.1)

$$e^{i\int d^4x V(\phi)} = \sum_n \left\{ \frac{[i\int d^4x V(\phi)]^n}{n!} \right\}.$$

But since $V(\phi)$ is a polynomial in ϕ, we can get the interaction term by differentiating the path integral (3.2) to get (3.1)! Thus we rewrite (3.1)

$$\langle 0|0\rangle_I^j = N \int \mathscr{D}\phi\, e^{i\int d^4x\{\mathscr{L}_0 + j(x)\phi(x) + V(\phi)\}}$$

$$= N \int \mathscr{D}\phi \sum_n \left[\frac{(i\int d^4x\, V(\phi))^n}{n!}\right] e^{i\int d^4x\{\mathscr{L}_0 + j(x)\phi(x)\}}$$

$$= \sum_n \frac{\left(i\int d^4x'\, V\left(\frac{\delta}{\delta j(x')}\right)\right)^n}{n!} N \int \mathscr{D}\phi\, e^{i\int d^4x\{\mathscr{L}_0 + j(x)\phi(x)\}}.$$

The beauty of this move is that now we can use our solution of $\langle 0|0\rangle_f^j$, for by obtaining the interaction term by differentiation we remove it from the path integral. We can thus write, using (3.3)

$$\langle 0|0\rangle_I^j = \langle 0|0\rangle_f^0 \sum_n \left\{\frac{\left[\dfrac{i\int d^4x'\, \lambda\delta^k}{\delta j(x')\ldots\delta(x')}\right]^n}{n!}\right\} e^{-i/2\int d^4x_1\, d^4x_2\, j(x_1)\Delta_F j(x_2)}.$$

For $\lambda\phi^k = V(\phi)$, and $\langle 0|0\rangle_f^0$ does not contain J. And finally, we get the time-ordered product we started with by

$$\langle 0|T(\phi_1\phi_2\phi_3\phi_4)|0\rangle_T = \left\{\langle 0|0\rangle_f^0 \frac{\delta^4}{\delta j(y_1)\delta j(y_2)\delta j(y_3)\delta j(y_4)}\right.$$

$$\left.\sum_n \left|\frac{\left(\dfrac{i\int d^4x'\, \lambda\delta^k}{\delta j(x')\delta j(x')\ldots}\right)^n}{n!}\right| e^{-\frac{1}{2}\int d^4x_1\, d^4x_2\, j(x_1)\Delta_F j(x_2)}\right\}_{/j=0} \quad (3.4)$$

This expression gives us the time-ordered product $\langle 0|T(\phi_1\phi_2\phi_3\phi_4)|0\rangle$ (the exact four-point function) in terms of a series in powers of the coupling constant. It is analogous to our earlier expression $\sum_k \langle f|S_k|i\rangle$ for the transition amplitude $\langle f|S|i\rangle$ and similarly generates the Feynman diagram expansion of, in this case, $\langle 0|T(\phi_1\phi_2\phi_3\phi_4)|0\rangle$. To see how this works, we will look at the second-order contribution to $\langle 0|T(\phi_1\phi_2\phi_3\phi_4|0\rangle$ in ϕ^4 theory. Then $k=4$, $n=2$ and we have $4+2.4=12$ $\frac{\delta}{\delta j}$'s operating. Looking at the propagator term and expanding it in a powers series,

$$\sum_m \frac{\{-\tfrac{1}{2}\int d^4x_1 d^4x_2 \, J(x_1)\Delta_f(x_1 x_2)J(x_2)\}^m}{m!},$$

only the $m=6$ term will contribute, because only that term has twelve Js, all of which will be killed by the twelve derivatives. Any less J's and we get zero by our differentiating, any more J's, and we get zero when we set $J=0$.

Now the

$$\frac{\left[i\int d^4x'\lambda\left(\frac{\delta}{\delta j(x')}\right)^4\right]^2}{2!}$$

creates two vertices, each with four propagator ends attached. On the other hand

$$\frac{\delta^4}{\delta j(y_1)\ldots\delta j(y_4)},$$

gives us four free propagator ends. This is consistent with the two types of diagrams shown in Fig. 3.8. And the Feynman rules emerge with $(i\lambda)$ to each vertex, a propagator to each •——• and I am ignoring

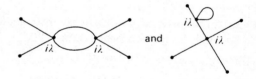

FIG. 3.8

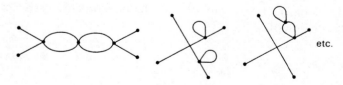

FIG. 3.9

counting factors. This gives us the second-order contribution to the four-point function. The third-order contribution will include terms like those in Fig. 3.9, while to the fourth order we have those shown in Fig. 3.10. And to get the S-matrix amplitude up to the mth order, sum all the relevant diagrams up to that order, remove external propagators, and multiply by the external wave functions (and some numerical factors).

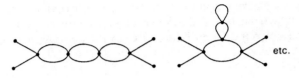

FIG. 3.10

For simplicity, we have used the self-interacting scalar field. But the case of different fields interacting with one another can be handled in a similar fashion. For example the QED vacuum-to-vacuum transition amplitude under the action of a source will be

$$\langle 0|0\rangle_I^{j,\rho} = N \int \mathcal{D}A_\mu \mathcal{D}\psi \mathcal{D}\bar{\psi} e^{i\int d^4x \{\mathcal{L}_0(\psi) + \mathcal{L}_0(A_\mu) + \mathcal{L}_{int} + j_\mu A^\mu + \bar{\rho}\psi + \bar{\psi}\rho\}}$$

$$\mathcal{L}_{int} \propto e\bar{\psi}\gamma^\mu A_\mu \psi$$

This can be handled as before because for $e = 0$, $\langle 0|0\rangle_I^{j,\rho}$ breaks up into

$$\langle 0|0\rangle_f^{j,\rho} = \langle 0|0\rangle_{f, A_\mu}^j \langle 0|0\rangle_{f, \psi}^\rho,$$

and each of these can be solved separately as before to get the electron propagator $S(x, y)$ and the photon propagator $D^{\mu\nu}(x, y)$,

$$\langle 0|0\rangle_{f, \psi}^\rho = \langle 0|0\rangle_{f, \psi}^0 e^{i\int d^4x \bar{\rho}(x) S(x, y) \rho(y) d^4y}.$$

$$\langle 0|0\rangle_{f, A_\mu}^j = \langle 0|0\rangle_{f, A_\mu}^0 e^{-1/2 \int d^4x d^4y j_\mu(x) D^{\mu\nu}(x, y) j_\nu(y)}.$$

And now we can get the interaction term as before by appropriate differentiations, and get our VEV of the time-ordered products by differentiation of

$$\langle 0|0\rangle_I^{j,\rho} = N \langle 0|0\rangle_{f, \psi}^0 \langle 0|0\rangle_{f, A_\mu}^0 \sum_n \frac{\left(i\int d^4x \left(ie \frac{\delta}{\delta\bar{\rho}(x)} \gamma^\mu \frac{\delta}{\delta j_\mu} \frac{\delta}{\delta\rho(x)}\right)\right)^n}{n!}$$

$$\times e^{i\int d^4x \bar{\rho}(x) S(x, y) \rho(y) d^4y} e^{-\frac{1}{2}\int d^4x d^4y j_\mu(x) D^{\mu\nu}(x, y) j_\nu(y)}$$

Virtual Particles and Quantum Field Theory 53

From this expression we see that in the perturbation expansion of, say, $\langle 0| T(\psi\psi\bar\psi\bar\psi)|0\rangle$ for the case of electron–electron scattering considered earlier, we get the familiar vertex factor ie and the fermion and photon propagators of the standard Feynman rules of qed.

We have obtained the Feynman diagram expansion of the transition amplitudes (via the expansion of the n-pt functions) without any mention of annihilation and creation operators. That eliminates one temptation to think that virtual particles mediate elementary particle interactions. But, of course, another temptation remains, namely the propagator.

Thus, for simplicity, consider the non-relativistic propagator $G(\mathbf{x}_2, t_2, \mathbf{x}_1 t_1)$ of the Schrödinger equation. Then for $t_2 > t_1$, we know that $\psi(t_2, \mathbf{x}_2) = \int d^3 x_1 \, G(\mathbf{x}_2, t_2, \mathbf{x}_1, t_1) \psi(t_1, \mathbf{x}_1)$.

Interpreting the wave function $\psi(t_i \mathbf{x}_i)$ as the amplitude for a particle to be found at $(t_2, \mathbf{x}_2)$, we have the natural interpretation of $G(\mathbf{x}_2 t_2, \mathbf{x}_1 t_1)$ as the amplitude for a particle at $(t_1, \mathbf{x}_1)$ to be found at $(t_2, \mathbf{x}_2)$. Now if $G(\mathbf{x}_2 t_2, \mathbf{x}_1 t_1)$ occurs in an expression does that mean it describes a particle going from $(\mathbf{x}_1 t_1)$ to $(\mathbf{x}_2 t_2)$? On the standard interpretation of quantum mechanics, of course not. As a special case of this, consider the familiar two-slit experiment. And suppose $\psi(\mathbf{x}_1 t)$ is the state of a particle at the slits at time t. Then the state at the point $\mathbf{y}$ on the screen at t' is (non-relativistically)

$$\psi(\mathbf{y}, t) = \int G(\mathbf{y}, t', \mathbf{x}, t) \psi(\mathbf{x}, t) d^3 x.$$

And supposing that at t, $\psi(\mathbf{x}, t) = C_1 \delta(\mathbf{x} - \mathbf{x}_1) + C_2 \delta(\mathbf{x} - \mathbf{x}_2)$ the state at t' at $\mathbf{y}$ is just

$$\psi(\mathbf{y}, t') = C_1 G(\mathbf{y}, t', \mathbf{x}_1, t) + C_2 G(\mathbf{y}, t', \mathbf{x}_2, t).$$

Since $G(\mathbf{y}, t', \mathbf{x}_1, t)$ is the amplitude for a particle at $\mathbf{x}_1$ at t to be found at $\mathbf{y}, t$, and similarly for $G(\mathbf{y}, t', \mathbf{x}, t)$, we know that, here, they cannot represent, respectively, a particle going from x_1 to y and one going from x_2 to y. And this is because, according to the standard interpretation, $\psi(y, t')$ is a superposition of $G(y, t', x_1, t)$ and $G(yt'x_2 t)$. Here, even though we do not have a superposition with respect to particle type or particle number, we still have a problem understanding how the electron gets from the source to the screen.

Returning to the Feynman diagram expansion (3.4) of the four-point function, we have a superposition of free propagator terms, just as we did in $\sum_k \langle f|S_k|i\rangle$, only we obtained this result using the path integral formalism, rather than with the operator formalism. So the superposition problem concerning particle number and particle type remain (they remain for quantum electrodynamics—for ϕ^4 theory we only have superpositions with respect to particle number). What I hope to have made clear, is that the question of the existence of virtual particles is just the question of the interpretation of the single-particle greens function—a question which we have already seen arises over the interpretation of the two-slit experiment within first-quantized non-relativistic quantum mechanics. So we reach the conclusion that we did before, and for the same reasons. We do not have good grounds for thinking that Feynman diagrams picture real, albeit 'virtual', physical processes by means of which elementary particles interact.

III

Although I have been arguing that virtual particles do not exist, I none the less think there is a particular kind of virtual process that we ought to take a closer look at. These are the ones that involve the so-called Fayddeev–Popov ghost fields. As a beginning, let us see where these fields come from and how they get into Feynman diagrams.

The ghost fields enter the picture when we fix the gauge of the Lagrangian of a gauge field A_u^a with, for example, a term of the form $-\frac{1}{2\alpha}(\delta_u A^{u,a})^2$. In that case the effective lagrangian, $\mathscr{L}$, becomes

$$\mathscr{L} = -\frac{1}{4} F^{uv,a} F_{uv,a} - \frac{1}{2\alpha}(\partial_u A^{u,a})^2 - \eta^a(x) \partial_u D^u W^a(x)$$

where, under an infinitismal gauge transformation δA_u^a,

$$\delta A_u^a = \frac{1}{g} \lambda D_u W^a(x)$$

$$D_u W^a(x) = \partial_u W^a(x) + gf^{abc} A_u^b W^c(x)$$

and $n^a(x)$, $W^a(x)$ are the ghost fields—scalar anticommuting fields which violate the spin-statistics theorem. We can get a feel for the origin of the last term by noting that while the 'classical' Lagrangian $-\frac{1}{4} F^{uv,a} F_{uv,a}$ is invariant under δA_u^a (since $F_{uv}^a = \partial_u A_v^a - \partial_v A_u^a$), the gauge fixing term $(\partial_u A^{u,a})^2$ is not. However, under $\delta A^{u,a}$, $(\partial_u A^{u,a})^2 \mapsto 2(\partial_u A^{u,a}) \partial_u \delta A^{u,a}$, so if $n^a(x)$ and $W^a(x)$ transform appropriately under the gauge transformation, the last term cancels the change in $(\partial_u A^{u,a})^2$, thus restoring a kind of gauge invariance to $\mathcal{L}$—the BRS invariance.[2]

To see why we need fermion fields, when we began with a purely bosonic field $A^{u,a}$, note that when we fix the gauge in the path integral

$$\int \mathcal{D} A^{u,a} e^{i \int d^4 x (-\frac{1}{4} F^{uv,a} F_{uv,a})}$$

by the Fayddeev–Popov method, we get

$$\int \mathcal{D} A^{u,a} \operatorname{Det} M \exp \left\{ i \int d^4 x \left(-\frac{1}{4} F^{uv,a} F_{uv,a} - \frac{1}{2\alpha} (\partial_u A^{u,a})^2 \right) \right\},$$

where $\operatorname{Det} M = \operatorname{Det} \partial_u D^u$ in the Lorentz gauge. Then by using the functional version of the identity

$$\operatorname{Det} M = \int \Pi^a dn^a \Pi^b dw^b e^{i n^a m^{ab} w^b},$$

for anticommuting variables n_a, w^b, we promote the $\partial_u D^u$ factor up into the exponent within the path integral.

For the connection of the ghost fields with virtual particles, note again that $D^u W^a(x) = \partial^u W^a + gf^{abc} A^{u,b} W^c(x)$ so we have the ghost-gauge field coupling, $n^a gf^{abc} \partial^u (A_u^b W^c)$ in our Lagrangian. In the Feynman diagram expansion, we represent this coupling by the vertex shown in Fig. 3.11. This is in addition to the three and four gluon–gluon vertices (Fig. 3.12), that come from $F^{uv,a} F_{uv,a}$. For example, then, the amplitude for two-gluon scattering will be a sum not only of pure gluon diagrams such as those in Fig. 3.13, but also diagrams containing internal ghost loops as in Fig. 3.14. Because the ghost fields couple to the gauge field, we must include the ghost field

---

[2] For an intuitive discussion of BRS invariance see I. J. R. Aitchison, *An Informal Introduction to Gauge Field Theories* (Cambridge: CUP, 1982).

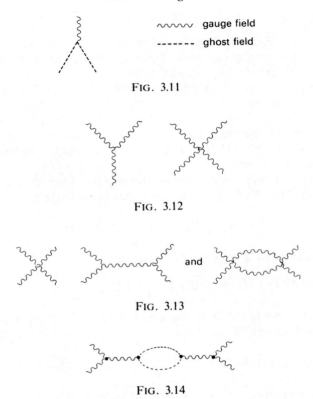

Fig. 3.11

Fig. 3.12

Fig. 3.13

Fig. 3.14

terms when we sum up the relevant Feynman diagrams in a calculation of an amplitude of the form $\langle \text{gluon} |S| \text{gluon}\rangle$.

Now it is sometimes said that since we only have to use ghost fields as internal lines in Feynman diagrams, and not as external lines, we do not have to worry about the fact that they violate the spin-statistics theorem. What is meant, I think, is that even if there are virtual 'ghost' quanta which violate the spin-statistics theorem, we will never observe the theorem being violated, so everything is in order.

But there is another problem with virtual ghost particles, which we can see if we consider a gauge, such as the axial gauge, in which the Fayddeev–Popov determinant, Det M, does not contain A_μ^a. In that case the ghost term $n^a(x) M^{ab} W^b(x)$ in the Lagrangian does not contain A_μ^a and there is no ghost field–gauge field vertex. Con-

Virtual Particles and Quantum Field Theory

sequently, we do not have to include any ghost loops in our Feynman diagram expansion of ⟨gluon |S| gluon⟩. Thus, whether the ghost fields are coupled to the gauge field depends on the gauge we choose to work in. It is just that it is more convenient to work in the Lorentz gauge with the gauge field–ghost field vertex because the gauge field and ghost field propagators are much simpler than the gauge field propagator in the axial gauge. So the virtual 'processes' which appear in the Feynman diagram expansion of ⟨glucon |S| glucon⟩ in the Lorentz gauge are just an artefact of our notation in a way that the other virtual 'processes' are not, for the gluon–gluon (or electron–photon) vertex cannot be transformed away.

By way of comparison with the ghost fields, we can look at the status of the vector potential in classical electrodynamics. It is often said in textbooks that because the magnetic field $\mathbf{B} = \nabla \times \mathbf{A}$ is invariant under a gauge transformation $\mathbf{A} \to \mathbf{A} + \nabla f(x)$, $\mathbf{B}$ is a 'real' physical field while $\mathbf{A}$ is just a calculational device. However, if we consider the two-slit experiment again, but this time with a long thin solenoid between the slits (see Fig. 3.15), the amplitude for a particle to go from the source to the screen will be

$$\langle x_1 t_2 | x_1 t_1 \rangle = \int \mathscr{D} x(t) \exp\left\{\frac{i}{\hbar} \int_{t_1}^{t_2} L_p dt + \frac{ie}{\hbar} \int_{t_1}^{t_2} \mathbf{V} \cdot \mathbf{A} dt\right\},$$

where L_p is the 'mechanical' Lagrangian of the particle and $e\mathbf{V} \cdot \mathbf{A}$ is the interaction Lagrangian of the particle with the electromagnetic field. Assuming $\mathbf{A}$ is not a function of time, so $\int_{\text{path } i} \mathbf{V} \cdot \mathbf{A} dt = \int_{\text{path } i} d\mathbf{s} \cdot \mathbf{A}$, the solenoid affects the interference pattern if $\oint_{I+II} d\mathbf{S} \cdot \mathbf{\bar{A}} \neq 0$. And this is the case for the solenoid even though $\mathbf{E}$ and $\mathbf{B}$ are zero everywhere outside of the solenoid.

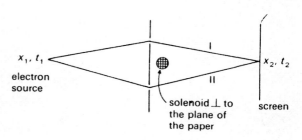

Fig. 3.15

But how can this be if **E** and **B** are the physical (real) fields, while **A**, which is non-zero outside the solenoid, is just a calculational device? Well, we could say that since

$$\oint_{\mathrm{I+II}} d\mathbf{s}\cdot\mathbf{A} = \int_{\substack{\text{surface}\\ \text{enclosed by I+II}}} d\mathbf{a}\cdot\nabla\times\mathbf{A} = \int_{\text{surface}} d\mathbf{a}\cdot\mathbf{B},$$

this is a non-local effect of the **B** field inside the solenoid. However, we can, alternatively hold that in quantum mechanics, this is a local effect of **A**, which we now conceive as a physical field. Does this make any sense in light of the fact that **A** is not invariant under a gauge transformation? It does, because while an arbitrary **A** can always be transformed to zero at any given point, it cannot, as we have seen, be transformed to zero along an arbitrary closed curve (or finite area). Thus, some of the degrees of freedom of **A** depend on our notation—on our choice of gauge, but not all do.[3] So, while in classical electrodynamics **A** can be regarded as a calculational device, it is able to take a physical significance when a new theory like quantum mechanics comes along. On the other hand, because it can be completely gauged away, the gauge–ghost field coupling, unlike the **A** field, is purely a result of our notation.

---

[3] Geometrically, this means that we cannot make the connection A_μ zero throughout a non-simply connected region R, even if the curvature $F_{\mu\nu}$ is zero throughout R.

4
Parsing the Amplitudes
ROM HARRÉ

INTRODUCTION

There are a number of reasons why a family of theories of such depth and sophistication as quantum field theory should be of interest to philosophers. In this chapter I want to bring out those points of interest that are of more general moment than the technical puzzles open only to those with, so to say, a professional grasp of the intimate details of this theory family.

As a near-fundamental theory of all physical processes QFT stands near the base of all explanatory hierarchies in the natural sciences, since it deals with the most ubiquitous processes of interaction between the most fundamental states of matter. I say 'near the base' since theories of the vacuum, recently elegantly summarized by Aitchison (1986), may turn out to be deeper than QFT in the sense that the processes described in the current conceptual repertoire of QFT may turn out to be mediated by processes in the vacuum itself. Metaphysicians can hardly fail to be attracted towards an exploration of the conceptual structure of an account of nature that purports to describe the world at such depth.

Contemporary physics has been the spawning-ground of a great many novel concepts, judged by reference to the kind of thing that the dominance of the Newtonian world-picture had accustomed us to. QFT contains all those concepts from classical quantum theory that have provided fodder for philosophers of physics for the last half century. But it also introduces in the idea of the virtual particle, a putative state of being which has to be dealt with by anyone committed to a generally realist reading of fundamental physical theories. Three issues seem to me to emerge from reflection on the sources and uses of this idea.

The concept of the 'virtual particle' appears along with a powerful iconic style of representation, the Feynman diagram. This way of

© Rom Harré 1987.

representing the physical meaning of amplitude terms together with a strong tendency towards a particulate way of interpreting the components of the mathematical descriptions of basic interactions, namely through what Aitchison has called 'photonic talk', leads to a drift towards a corpuscularian image of nature. I shall argue that this drift favouring particle expressions is not a mere convention of convenience, but reflects a deep feature of the nature of physical science, its dependence on experimental apparatus. Feyman diagrams use a visual image through which the mathematical representations of exchange processes are visualized. Since A. I. Miller's interesting study (Miller 1985) of the role of visualizability in fundamental physics one must be cautious in dismissing the image as merely heuristic. The status of that image and its main problematic component, the 'virtual' particle, is a fit subject for philosophical reflection. Cartwright's recent book (Cartwright 1983) has drawn our attention once again to the caution that is needed in interpreting concrete models of physical processes and their relationship to the mathematical expressions of a theory. I believe that philosophical reflection on these matters leads one to favour a generally Bohrian treatment of physics. The kind of dispositions that QFT leads us to want to attribute to the basic stuff are affordances, dispositions whose content is derived from the kinds of response they bring about in man-made apparatus. Finally it is worth noticing that the relationship between choice of symmetry group and introduction of quantum numbers in describing interactions differs in a very fundamental way from the relationship between the Lorentz group and relativity. In the latter the choice of symmetry group can be independently motived by reflection on the outcome of the Michelson–Morley experiment. So far as I can see in the former symmetries and conservations are mutually motivated. And this has important epistemological consequences. In this chapter I shall confine my remarks to the first and second of these issues, since to go into the third in detail would involve little more than a repetition of an argument fully worked out elsewhere (Harré 1986).

There is a third aspect of QFT that strikes me as being of general philosophical interest. Looked at retrospectively QFT begins to shape up as a research programme. While I do not believe that its earliest phases of development really had that degree of orderliness, nevertheless the latter part of the history of these developments does seem to have the kind of programmatic structure that makes it worthwhile to ask whether there is some characteristic pattern of reasoning to be discerned in it. I think that there is. There are two striking features about QFT as a research programme. It is dominated conceptually by the existence and popularity of a certain kind of *experiment*, and this will come out in the discussion of the conceptual innovations of QFT. But it is also structured by a pattern of reasoning that involves the interplay of a system of analogies. Analogical reasoning is organized by asymmetries between sources and subjects of analogy. In QFT the components of analogy keep shifting from source to subject and back again. It is these transformations of role that seem to me to underlie the dynamics of thought in the developing research programme of QFT.

THE MECHANICS OF CONCEPT CREATION

The Feynman diagrams, through which QFT is thought concretely as a physical theory, are an obvious locus for the insights of both Miller and Cartwright. From Miller's point of view there is something significant about the iconic force of these images despite Feynman's warnings about their readability as representations of anything physical. I want to try to bring out the significance of the particular kind of iconicity these diagrams display. This line of thought favours a qualified realist reading of the diagrams. On the other hand, from Cartwright's standpoint, in which models are seen as often detached from the reality they represent, Feynman diagrams are a fine example of a class of image which should not be confused with a representation. Should we not leave it at that? In particular is there anything useful that can be said by philosophers, particularly those such as myself, whose tenuous grasp of this highly technical family of theories is at best vicarious?

Reflection on the nature of Feynman diagrams can yield deeper insights than merely noticing a human psychological imperative to use iconic modes of representation. I believe that these diagrams illustrate the power of the material practices of a scientific community—those involved in doing experimental work—to shape concepts in the basic theory. I mean 'work' in the sense of shaping material stuff in certain quite definite ways—ways that determine which aspects of the enormous range of possible effects that natural processes might have on human artefacts will actually become available to human observers. We have no right to suppose that our apparatus stands in some kind of privileged relation to what natural processes can do to bits of material stuff. I want to try to support that idea that the exigencies of experimental procedures in high-energy physics are as much responsible for the usefulness of the iconic properties of Feynman diagrams, and of their epistemic power, as any alleged feature of human cognition. I do not think that the best minds in the physical sciences are likely to be trapped at Piaget's stage of concrete operations! In this way I want to throw some light on the way talk of 'virtual particles' should be interpreted by philosophers.

We have to account for the fact that 'photonic' language seems to be entirely natural in interpreting the electromagnetic interaction, and at the same time, for the success of experimental research programmes which set out to 'find the W and Z particles' with a certain particular kind of apparatus, an apparatus whose end-product is *tracks*. As a philosopher my problem is to understand the significance of particle talk and of the icons that reflect it. Other vocabularies and other icons are on offer, at least in principle.

I want to argue that the conceptual structure of the idea of an intermediate vector particle is not that typical of an inspired hypothesis, a free creation of the human mind, such as Einstein said he favoured. The concepts of the family of IVPs are constrained in a philosophically interesting way. In the end these concepts are constrained by reference to the way experimental apparatus is constructed. To show this I shall follow a typically elegant exposition by Aitchison and Huey, which I shall quote in full. It is an example of the reasoning pattern I have called 'parsing the amplitude'. Faced with an algebraic formulation of a concept cluster physicists typically arrive at an interpretation by picking out those components for which a reasonably clear-cut meaning has already been given, sometimes through an empirical technique or more usually through a model. Any algebraic residue is dealt with *ad hoc*, by drawing on some further analogy. This is the way Maxwell reached the concept of 'displacement current' (cf. Whittaker 1951). There was a feature of the mathematical description of the electromagnetic field which did not have a direct interpretation within the framework of his existing conceptual resources, tied in as that was to the electromagnetic phenomena. By applying the ether *models* the interpretation of the 'mystery' term as displacement current falls out. It is one of the main ways the conceptual resources of physics grows. Can the same analysis be offered of how IVP concepts are born?

Parsing the Amplitudes

To illustrate the point I quote extensively from a particularly nicely laid out derivation of an IVP concept by Aitchison and Huey (1983, pp. 29–30).

"A Feynman diagram is a pictorial representation of a process, corresponding to a particular transition amplitude. We have derived the amplitude for πK electromagnetic scattering to lowest order in α. Our result is

$$\mathcal{A}_{\pi^+ K^+} = -iN_1 N_2 N_3 N_4 (2\pi)^4 \delta^4(p_3 + p_4 - p_1 - p_2) \quad (4.1)$$
$$\times e(p_1 + p_3)_\mu \frac{(-g^{\mu\nu})}{q^2} e(p_2 + p_4)_\nu.$$

A Feynman diagram is useful because the different pieces of this expression can be associated with different elements in the diagram. Furthermore, this association generalizes to higher-order calculations. Consider our example: the Feynman diagram for lowest-order $\pi^+ K^+$ electromagnetic scattering is drawn in Fig. 4.1. Let us comment on the factors that appear.

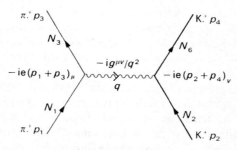

FIG. 4.1

1. There is a normalization factor N_i for each 'external' line in the diagram.
2. The interaction between the mesons is due to the exchange of a photon, represented by the wavy line (∿∿), which is associated with the factor

$$-ig^{\mu\nu}/q^2. \quad (4.2)$$

This is the so-called 'photon propagator': the indices μ and ν refer to the ends of the photon line and reflect the fact that the exchanged photon is a spin-1 particle. The factor q^2 is often referred to as 'the mass squared of the virtual photon'. This is because a real photon satisfies the free-field equation

$$\Box A^\mu = 0. \tag{4.3}$$

For plane-wave solutions of the form $e^{-iq\cdot x}$, the photon 4-momentum q must satisfy

$$q^2 = 0 \tag{4.4}$$

reflecting the fact that a real photon is massless. As we have seen, an interacting photon satisfies

$$\Box A^\mu = J^\mu_{em} \tag{4.5}$$

and consequently

$$q^2 \neq 0. \tag{4.6}$$

The interacting photon is said to be 'virtual', or 'off mass shell', and q^2 is termed the mass squared of the virtual photon."

The parsing of the amplitude expression (4.1) as the photon propagator and the construction of the associated Feynman diagram are two faces of the same conceptual step, alternative ways of representing the same pattern of reasoning. A glance at the structure shows that all the terms except $-g/q^2$ are fixed by the empirical description of the interaction as a 'spinless electron' scattering from a 'spinless muon'. The authors offer the reader this case merely by way of example, since the strong interaction between these particles dominates the pattern of any real-world scattering. Nevertheless the lines in the diagram are tantalizingly reminiscent of photographs or computer reconstructions of tracks. How is the meaning of the algebraic residue, $-g/q^2$ to be determined? Calling it the 'photon propagator' reflects the source of the interpretation as an IVP, and not anything else. There is a clear, formal analogy between a real photon expression and the expression in question. I believe that it is this analogy, and I suspect that it is to some extent contrived, that supports the introduction of further bits of 'photonic' talk. For the next step is to follow through with the observation that the term fits a spin-1 particle, and with a treatment of the fact that $q^2 = 0$, as it should for the 4-momentum of a 'normal' photon. The Feynman diagram is just an iconic alternative to this discursive reasoning. Looked at in the context of this pattern of reasoning these famous diagrams are important not for their heuristic, iconic properties, but because they are a way of parsing the amplitude. But not any old way of performing that important step. To the question 'Why read the "boson" term in a language of corpuscularian or particle concepts?' the answer suggested by the above analysis surely is 'Because we have already read the "fermion" terms in a vocabulary based on particle concepts.' But why read the fermion terms in such a vocabulary? Why not use de Broglie's rules to read them in wave concepts? The answer I suggest is that the apparatus we like to make produces *tracks*. Once we reach this point it is easy enough to see how that fact exerts its influence right back to the birth of the IVP.

A TURN TOWARDS BOHR'S PHILOSOPHY OF PHYSICS

In the above analysis the unravelling of the conceptual structure of the basic concepts of QFT has still left the reality question unresolved. Do photonic terms that appear in the QFT way of parsing amplitudes denote anything at all? If they do, is it anything like what a photon, as a referent, is supposed to be? It should be clear already that there cannot be a straight answer to either of these questions. If there is to be a realist reading it cannot be via a straightforward reification of the wiggly line in a Feynman diagram, however much the 'outer' lines look like tracks in photographs of subatomic 'sunbursts'. Another kind of concept must be added to our repertoire.

Cartwright's interesting and important book contains one fatal flaw. Her one-line dismissal of dispositional accounts of basic physical concepts simply will not do. I shall try to show that the intriguing concept of 'virtual particle' can be given a realist reading, not in terms of the Feynman icons, but through the use of dispositional concepts. The Feynman icons serve to control the importation of conservation principles and show with clarity the role played by the exigencies of experimental procedures in taking the mathematical representations of QFT one way rather than any other. They play an important heuristic role. But the new style of concept needed for the whole structure to make sense is the Gibsonian 'affordance'. The relevant affordances will be tied to the insight that whatever a physical being is, it must manifest some of its causal power in the behaviour of apparatus.

Freed from the canard of positivism Bohr's philosophy of science is once again on offer as the main opposition to Einsteinian realism. If there are any empiricists about these days they are keeping their heads down. A recent advance in our understanding of dispositional concepts, due to J. J. Gibson, has helped us to get clearer about Bohr's ideas. Expounding the ideas of Neils Bohr is a hazardous business, but armed with the scholarship of John Honner (1987) and the concept of an affordance I plunge in with two root ideas.

The first is the idea of apparatus as 'shaping up the "glub"'. The Kantian idea that the things, events, and causal relations of the phenomenal world are really structural properties imposed on a flux of sensation having their origin in a priori concepts seems to have influenced Bohr in his notion of a phenomenon. He borrows the structure of the Kantian account of experience for his analysis of physics. The ur-stuff of the physical world, which I shall nickname the 'glub', corresponds to the sensory flux. The pieces of equipment human beings construct correspond to the Kantian schematisms. Their states are phenomena, created by forcing the glub to manifest itself in ways that are predetermined by the structure and other properties of the equipment. A certain kind of apparatus shapes up the glub into particles, or better particulate phenomena. These particles—shaped up glub—can exist nowhere else but in relation to that kind of apparatus. What properties can we attribute to the glub on the basis of what happens in an experiment?

This problem existed in classical chemistry, but philosophers singularly failed to address it. As a child I spent all my pocket-money and most of my leisure hours on chemistry. I think I was nine when I contrived to save enough to buy a retort, a piece of equipment necessary to fulfil my ambition to isolate bromine. I watched the dark droplets coalesce on the walls of the retort with religious awe, and then drip into the container. I dare say liquid bromine exists nowhere in the universe. Yet a schoolboy was able to bring 'something' out of nature to within the light of human experience. But bromine liquid was not buried or concealed in nature. Fifty years later, thanks to the ideas of Bohr and Gibson, I have at last got a tenous grasp on the metaphysics of those long-ago events. Neither the apparatus nor free, liquid bromine exists in nature. The substance 'bromine' is as much an artefact as the retort. Retort and bromine exist as a reciprocal pair. The retort is shaped by the exigencies of condensing vapours, while liquid bromine is made possible, as a stuff, by the apparatus. Without knowing it I had achieved a Bohrian phenomenon.

It is a short step to complementarity. We humans can build, so far, only two main kinds of apparatus, each kind shaping up the glub in different ways, particulate ways and undulatory ways[1]. It is phenomena—the glub as shaped up by apparatus—that present themselves within the constraints of complementarity. But this move is not enough. A new kind of concept is needed to express what has turned up in phenomena as dependent in some causal way on the unknowable properties of glub. With affordances thought of in the manner of J. J. Gibson a great many difficulties can be resolved, and the level of our essential ignorance of nature nicely gauged.

Gibsonian affordances are a special case of the familiar concept of disposition. They can be used to create property-concepts which can be ascribed to a thing or substance on the basis of the way it behaves in certain well-defined conditions, though we are in ignorance of the intrinsic properties of that substance. By this device the nature of a substance is tied into the explanation of its behaviour prior to the discovery of its constitutive properties. The use of dispositional concepts has been characteristic of all branches of physics since the days of Gilbert's *De magnete*. But an important feature of dispositional concepts remained unnoticed until the 1950s when Gibson began to formulate his now famous theory of perception. The way the state of things which is the underlying basis of a causal power is manifested depends on the kind of being it is manifested to. A bee sees a rose differently from the way we might. Bees are sensitive to light of much shorter wavelengths than we can detect visually. So a garden affords a different chromatic topography to bee vision from that which it affords to human beings. What a physical system *affords* is relative to the nature of the being which interacts with it, in particular what states it is capable of taking up. Affordances are dispositions of physical things relativized to that with which they interact. For complex physical things like roses we are able to say how its one physico-chemical constitution allows us to see it as red, and a bee to see it as a purplish hue with its ultraviolet skewed sensibility. But suppose the things or stuff we are trying to study is of an unknown constitution. Then all we do is to ascribe whatever bouquet of affordances to it we can. The glub is such a stuff. Complementarity, expressed in affordance terms, goes like this: the glub affords particulate phenomena for one species of apparatus and undulatory for another, and human beings cannot make this equipment so that one of each kind can occupy the same place at the same time. Pieces of apparatus are themselves glub, shaped up as wires, metres, and maybe Bohr's favourite nuts and bolts, by human perceptual apparatus and the rest of the material universe. This is a realist theory of science because the glub really does possess the affordances to be so shaped up. An apparatus would not afford particulate 'clicks' unless the glub was of such a nature, whatever that is, to afford electron phenomena. But since the concept of any affordance includes an ineliminable reference to the equipment doing the shaping (Bohr's 'phenomenon' again), an affordance is not a detachable and simple property of the glub. Unless every conceivable style of apparatus would shape up the glub the same way we can never eliminate the style of apparatus from the physical description of the world. As a fundamental or near fundamental physical theory QFT must deal in affordances. It is my contention that 'virtual particle' talk is affordance talk, the 'particle' aspect of that talk coming from the exigencies of the way experiments are done in high-energy physics. It is realist epistemologically, but it does not license any loose thoughts about there being really little bits of energy-stuff exchanged in a electron–muon interaction. The iconic form of the Feynman diagrams is important, since it is not just a convenient model or convention of representation. It is reflection of the control exercised over affordances by the fact that our preferred apparatus squeezes the glub into such a form that it affords tracks.

PATTERNS OF REASONING

Philosophers have a professional interest in trying to extract the patterns of reasoning that are inherent in the practice of a physical science. One can detect an interesting interplay between generic and specific concepts in the broad structure of QFT particularly since it has developed into a conscious research programme. In the pattern of reasoning described above that I have called 'parsing the amplitude', the upshot of the use of the photon analogy for interpreting the 'spare' term in the expansion of the amplitudes is the creation of a genus, the 'photonics'. This genus has two species, real and virtual. The metaphysical status of the virtual species remains to be determined. The real photon is manifested in luminiferous phenomena, the virtual photon is invoked in theory. All this would be mildly interesting, but scarcely remarkable but for the research programme initiated by the QFT treatment of the weak interaction. The above schema of parsing by analogy and then setting up a genus with two species is again applied, but in reverse. The W particle emerges as a species in a parsing operation which starts with the virtual member of the species pair under the genus. Its properties are worked out, so far as I understand the physics, by the application of such general principles of mathematical epistemology as conservation of quantum numbers, local gauge invariance, and so on. Then the concept of the 'real' species is created by running the pattern of reasoning originally invoked in electromagnetic QFT backwards. It is the concept of the virtual species that is legitimated via its explanatory power in the tidy accounting of energy budgets and quantum number conservations of the weak interaction. The programme of experimental activities that is to be called 'the search for the W particle' is controlled by the concept of the real species. That concept is dependent on an analogy in which the concept of the virtual species is the source and the real species the subject; just the reverse of the way the programme of QFT developed in electromagnetic theory.

What could lie behind this pattern other than heuristics? Well, in the end, an experimental programme must be mounted. In the usual form that project realism takes in physics and other natural sciences, the conceptual structure of the theoretical entity invoked in explanations specifies a research programme. If genes are bits of chromosomes we know where to start looking for them. But IVPs are not candidates for experimental disclosure, only their 'real' counterparts. So I suggest that the exigencies of experimentation, the fact that we are practically confined to looking for *tracks*, fixes one end of the reasoning pattern, namely how the real species must be conceived. Now I think I can see an overall structure to the conceptual basis of weak interaction physics. Photonic ways of talking are not just reflections of the convenience and elegance of the Feynman icon for parsing amplitudes, but also reflect the exigencies of experimental programmes. But in this case we substantiate the genus not by finding examples of the species of the very entity proposed in the *pictured* theory, but one which, though of the same genus, belongs to a different species. Finally, it is worth noticing that all the IVPs on show are species of genera which belong to a higher-order photonic family.

The general theory of affordances ties physical concepts tightly to the possibilities of experimental procedures and only loosely to the occurrent states of the glub. I take the Bohr lesson to be that there may be no way assigning occurrent states to the glub. But the advent of vacuum physics seems to raise the issue of occurrent states again, and rather changes the ontology implicit in QFT. I am greatly indebted to an excellent paper by Ian Aitchison in which he brings together current developments in theory of the vacuum. If we imagine an oscillator plenum—a field of oscillators—the vacuum corresponds to a condition in which the average energy of the oscillators is zero. This permits non-zero fluctuations to be imagined. All this is very familiar no doubt to readers of this book. From the point of view of a philosopher this story looks like the familiar progress in theorizing in which certain dispositions are grounded in some state of an underlying substance. The oscillator plenum plays the role of substance and its states of excitation the occurrent groundings. Photons are a particulate representation of quanta of excitation energy of such oscillators. Aitchison points out that the Lamb shift, for instance, explicated in terms of the oscillator plenum, is capable of a photonic description, namely as due to the emission and reabsorption of a virtual photon. Now photonic talk is motivated somewhat differently from the patterns of reasoning I have sketched above. In terms of the properties of the oscillator plenum photons are elementary quanta of excitation of elementary oscillators. So fluctuations in the state of those oscillators can be given a photonic description and hence the Lamb shift can be described in terms of an IVP. As I understand it this kind of explanatory pattern is quite general in QFT.

The naïve philosopher's question: 'Are IVPs real, though virtual?' now takes on a different complexion. Since photonic talk is a convenient way of describing events in the oscillator plenum, the realist tack then shifts to the asking of metaphysical questions about that plenum. And that issue will, I hope, be among those to be addressed at our next colloquium, convened around the physics of the vacuum.

It seems to me that the marked success of photonic *talk* does not support the hypothesis that the world is corpuscularian. But to go as far as Weingard has put it in conversation, that virtual particle talk is nonsense, though he used a bovine metaphor, will not do either. The photon genus is not just a heuristic image allowing us to think concretely in terms of spin-1 particles; it is also a constraint that ensures that experimental programmes worked out from the theory conform to the exigencies of current empirical practices.

To summarize my argument: I want to claim that the real/virtual distinction is a substantial one, with physical and not just heuristic significance. A real particle is the referent of an expression for which the Bohrian affordance in an experiment motivated by that expression and the others which go with it, is a track. A virtual particle is the referent of an expression, analogous to that for a real particle, for which, could we construct a track-yielding apparatus, would afford a track or tracks. How do we know that the counterfactual is plausible? The answer is built into the analysis already. It is because we have so parsed the amplitude that it yields a 'spare' term which is neatly interpreted using a photonic analogy. It's all in the family, so to speak.

[1] I owe to Simon Saunders the observation that apparatus which shows that the 'glub' affords waves, if it does, is not on all fours with that which creates tracks. An *array* of Geiger counters still detects phenomena click by click. This asymmetry deserves more investigation.

REFERENCES

AITCHISON, I. J. R. (1985): 'Nothing's Plenty: the Vacuum in Modern Quantum Theory', *Contemporary Physics* **26**, 333–91.
—— and HUY A. J. C. (1982): *Gauge Theories in Particle Physics* (Bristol: Adam Hilger).
BOHR, N. (1963): *Essays (1958–64) of Atomic Physics and Human Knowledge* (New York: Wiley).
CARTWRIGHT, N. (1983): *How the Laws of Physics Lie* (Oxford: Clarendon Press).
GIBSON, J. J. (1979): *The Ecological Approach to Visual Perception* (Boston, Mass: Houshton Mifflin).
HONNER, J. (1987): *The Description of Nature: Niels Bohr and the Philosophy of Quantum Physics*. Oxford: Oxford University Press.
MILLER, A. I. (1984): *Imagery in Scientific Thought* (Boston, Mass.: Birkhauser).
WHITTACKER, E. (1951): *A History of Theories of Aether and Electricity* (New York: Harper & Row), 250–1.

5
Three Problems of Renormalization
PAUL TELLER

1. INTRODUCTION

What is renormalization in quantum field theory? Popular descriptions talk about 'throwing away infinities' which occur in calculations, leaving behind numbers which agree extraordinarily well with experiment. This sounds crazy. But the non-technician has a hard time evaluating the situation because heretofore all presentations of renormalization have been buried in an impenetrable mass of calculational detail.

In this paper I aim to strip away the physics so non-experts can see the logic of the renormalization procedure. When laid bare, we will find that the procedure does leave us with interesting questions, but that it need not partake of the mathematical absurdity suggested by unargued talk of 'discarding infinities'. I will also survey a range of attitudes towards renormalization expressed by the physics community. While in important respects all agree as to how renormalization should be carried out, one can discern three different attitudes towards the procedure.

This paper covers the central facets of renormalization as the procedure was understood around 1950. I will do no more than mention important developments of the last two decades dealing with the renormalization group and application to non-abelian gauge theories. As far as I can determine from my very limited understanding of these newer topics, they concern fundamentally new concepts central to physics' description of matter, but do not raise new methodological problems comparable to the spectre of 'discarding infinities'. I should mention that this presentation overlaps with another I am preparing, to be entitled 'Infinite Renormalization'. The

---

© Paul Teller 1987. So many have helped me with this work that it would be silly for me to try to remember who they all were and thank them here. However, Michael Redhead and Gordon Fleming have contributed so much to what I have been able to understand that I hope others will forgive me if I single these two out for special thanks. This work was generously supported by NSF grant No. SES-8217092.

present paper summarizes the basic ideas of renormalization which I discuss in somewhat more detail in the companion article and then extends the presentation in examining further aspects of renormalization methods and problems.

2. THE FIRST PROBLEM: INFINITIES IN THE THEORY

Quantum field theory presents equations describing the interaction between different quantum fields. But no one begins to know how to solve these equations exactly. One can do a lot with a certain theoretical quantity, S, which figures in the evaluation of many different observable quantities. Since the physical meaning of S, called the 'propagator', will not be important, let us just think of S as a useful theoretical quantity and focus all our attention on how physicists try to get approximate values of S which they can use in calculating experimental results.

S describes aspects of a particle, say an electron—I will have the example of quantum electrodynamics foremost in mind, though everything I will say applies, at least in outline, to renormalization in other theories. All we need to know about S is that it is a function of the particle's mass, which we indicate by writing $S(m)$. Physicists proceed by calculating first, second, third and higher approximations to $S(m)$, which we write as $S_1(m)$, $S_2(m)$, $S_3(m)$, ... Of course the hope is that these approximations become more and more accurate estimates of the value of $S(m)$.

No problems arise with $S_1(m)$, which already yields quite accurate results. But evaluation of $S_2(m)$ leads to trouble. The problem arises because $S_2(m)$ must take into account the particle's interaction with its own field. Adding up all the bits of this self-interaction gives an integral, $\int_0^\infty g(k)dk$, which is infinite. This is the first of the 'infinities' which are 'thrown away'.

To study this problem, let us focus on what one more exactly means by an infinite integral. An integral with an infinite upper limit of integration is, properly speaking, the limit $\int_0^\infty g(k)dk = \lim_{L\to\infty} \int_0^L g(k)dk$. To say that this integral is infinite is to say that $\int_0^L g(k)dk$ increases without bound as $L\to\infty$, in other words, that $\int_0^L g(k)dk$ diverges as $L\to\infty$. Speaking exactly we should describe $\int_0^\infty g(k)dk$ as divergent.

If we insist on looking at the $L\to\infty$ limit, without discussing comparisons between quantities as we go to the limit, our approximation $S_2(m)$ indeed makes no sense. But we do not have to do that. Let us instead study the situation when L has some large but finite value, that is let us see how calculations proceed using $\int_0^L g(k) dk$ instead of $\int_0^\infty g(k) dk$, for some large but finite value of L. L is called a *cut-off* (we have 'cut off' the divergent top of the integral), and in making the divergent integral finite by putting in a cut-off we say that we have *regularized* the integral.

As long as we work with $\int_0^L g(k) dk$ we have only finite terms appearing in the calculation of $S_2(m)$, and we are free to manipulate them as we like. It turns out that we can group $\int_0^L g(k) dk$ with the mass, m, so that we can write $S_2(m)$ in the form $S_2(m - \int_0^L g(k) dk)$. Now comes the critical part of the argument. The divergent integral arose in the description of the particle's interaction with its own field. The fact that $\int_0^L g(k) dk$ can be grouped with the mass reflects the physical fact that the particle's self-interaction makes the particle act as if it had a mass which differs from the mass it would exhibit if there were no interaction. Let us use the description *bare mass* for the mass of the particle in the absence of interaction, which we write as m_0. Thus, if there were no interaction, if, as physicists like to say, we could 'turn the interaction off', our second approximation would be correctly written as $S_2(m_0)$. But we can never turn the interaction off. Thus the second approximation always takes the form $S_2(m_0 - \int_0^L g(k) dk)$.

Let us pay close attention to this last expression. Since we can never turn the interaction off, we can never observe the bare mass, m_0. We can only measure the mass which the particle presents to us, dressed in its interaction with its own field. That is we can only measure the mass quantity

$$m_r = m_0 - \int_0^L g(k) dk \tag{5.1}$$

which we call the renormalized mass. Since we can never turn the interaction off, we have to accept m_r as the real mass, the only mass which shows up directly or indirectly in experiment. And the value of this real mass is firmly fixed for us by experiment.

Finally, let us face up to the fact that we have described S_2 inaccurately because we have put in the cut-off, L, on the divergent integral $\int_0^\infty g(k) dk$. The point is that (5.1) is going to hold no matter how large L is. m_r is fixed by experiment, and so, as we let L get larger and larger, we must think of m_0 and $\int_0^L g(k) dk$ as differing by the experimentally fixed quantity, m_r. For each increasing value of L, the diverging quantity $\int_0^L g(k)\mathrm{d}k$ is 'absorbed' into the fixed, finite m_r, on which S_2 and all experimentally observable quantities depend. Thus as we go to the $L \to \infty$ limit all directly and indirectly observable quantities are finite. As a calculational scheme this procedure is perfectly consistent.

While this basic renormalization argument involves no mathematical inconsistency, we must acknowledge conceptual strain. As we proceed to the $L \to \infty$ limit we have to think of the bare mass, m_0, as being larger and larger, so that, as in (5.1) its difference with $\int_0^L g(k)\mathrm{d}k$ is always exactly the observed m_r. Conceptually, it still sounds as if, in some sense, we are told that there is an infinite bare mass balanced by an infinite amount of self-interaction, just so as to yield the observed m_r. I will have more to say about this puzzling situation. But first I want to describe the second problem of renormalization.

3. THE SECOND PROBLEM: FINITE CORRECTIONS

The renormalization procedure does more than absorb divergent quantities. Put crudely, from the divergent quantities the procedure also separates finite 'corrections' which, so to speak, get left behind after renormalization. On its face, this would seem to be absurd. From an infinite quantity one can separate out any finite amount you like. But, to add insult to injury, this seemingly absurd procedure produces absurdly good results, indeed, the most accurate predictions to be found anywhere in science.

Once again, attention to the argument's form dispels the impression of mathematical black magic. From the start, let us discuss the situation in terms of finite quantities by putting a cut-off, L, on the divergent integral. It turns out that the integrand can be split into two parts, one of which is multiplied by a parameter, q (called the 'impact parameter', which describes the amount of momentum transferred in an interaction):

$$\int_0^L g(k)\mathrm{d}k = \int_0^L (g_i(k) + q g_f(k))\mathrm{d}k \quad (5.2)$$
$$= \int_0^L g_i(k)\mathrm{d}k + q \int_0^L g_f(k)\mathrm{d}k$$
$$= I(L) + qF(L)$$

with $\lim_{L \to \infty} I(L) = \infty$, and $\lim_{L \to \infty} F(L)$ well defined.

Three Problems of Renormalization

The subscript 'i' on the first integrand indicates that when this integrand is integrated from 0 to ∞ it will diverge, and the subscript 'f' on the second integrand indicates that its integration from 0 to ∞ gives a well-defined result. '$I(L)$' simply abbreviates the first integral which will diverge in the $L \to \infty$ limit and '$F(L)$' abbreviate the second integral which has a well-defined L limit.

Let us put (5.2) into (5.1):

$$m_r = (m_0 - I(L)) - qF(L). \tag{5.3}$$

The crucial stage of this argument turns on noting that in the situations in which we conventionally measure the mass, for all practical purposes $q = 0$. Thus in these situations $m_r = m_0 - I(L)$. This is the value of m_r which we write down in our reference manual as the particle's mass. Now, suppose that we go on to measure some quantity in a situation in which q is significantly different from zero. In doing our calculations we use the value of the mass written down in the reference manual, the value calculated when $q = 0$. But in this situation the particle acts as if its mass were altered by $qF(L)$ from its $q = 0$ value. One must keep clearly in mind that, at this stage in the discussion, we work with the cut-off L, so that *all* quantities are finite. Accurate calculation requires that the $qF(L)$ contribution not be neglected, and indeed this term, and a similar one arising from the renormalization of charge, are precisely what are called 'radiative corrections'.

Now let us take the $L \to \infty$ limit. (5.3) holds for all values of L. Since we have decided to measure the mass, m_r, when $q = 0$ we have, for every value of L, $m_r = m_0 - I(L)$. But for *every* value of L, if we are concerned with calculating a quantity in a situation in which $q \neq 0$, we must add $qF(L)$ back into the $q = 0$ value of m_r. So for every value of L, $I(L)$ disappears from our calculations, and in the limit we are left only with well-behaved, finite quantities. In particular, $\lim_{L \to \infty} F(L)$ is well defined, and to be taken into account in any case in which q is significantly different from 0.

How should we understand the situation in which we have let L go to infinity? Mathematically the procedure is perfectly consistent, but it is not clear that the situation at the limit makes much conceptual sense. One might try here to unpack our thinking about the limit as an 'ideal point' properly to be understood in terms of all the cases that arise along the way to the limit. I have in mind here a strategy similar to the one we use in thinking about instantaneous velocity as a way of covering all the statements about the ratio of distance traversed to time elapsed, over smaller and smaller intervals. For each value of L consider the 'theory' $T(L)$ which we get by putting in L as a cut-off in all the divergent integrals in the original theory. We can now understand talk about the theory in the $L\to\infty$ limit as talk about the sequence of 'theories' that arise as the limit is taken. In particular, we can understand talk about the 'infinite' bare mass as the sequence of statements about the values of the bare mass, $m_0(L)$, in each of the theories, $T(L)$, and talk about the infinite value of $I(\infty)$ as talk about the divergent sequence of values of $I(L)$ in the theories $T(L)$. And we can understand the statement that the difference between the two 'infinite' quantities, m_0 and $I(\infty)$ is the finite quantity m_r as the statement that in each theory, $T(L)$, along the way, $m_0(L)-I(L)$ has the constant finite value m_r. Finally, if we move into a situation in which $q\neq 0$ we must, in each $T(L)$, take $qF(L)$ into account. But since $F(L)$ converges to a well-defined limit, for all values of L larger than some sufficiently large value we may neglect the difference between $qF(L)$ and $q\lim_{L\to\infty} F(L)$ and just use $q\lim_{L\to\infty} F(L)$ throughout.

A final comment before closing this section. What if we measure the mass at some value other than $q=0$? We are perfectly free to do this, and if so, our measured mass will differ from the conventional value. The value of the mass is not really a constant but depends on q as indicated in (5.3). Note that I have not specified the sign of $F(L)$, so I have not said whether $qF(L)$ makes m_r larger or smaller. In fact, different theories give different results in this regard. The same sort of thing is true of the electric charge and other 'coupling constants' in quantum field theory. This fact is what one refers to in talking about 'running coupling constants': in our case this just means that the value of the mass is a function of q. In the case of quantum chromodynamics (QCD) the coupling constant actually becomes smaller as q increases, resulting in quarks behaving as if they were free particles when they are very close together ('asymtotic freedom'). It is also believed that the QCD coupling constant becomes larger without limit as $q\to 0$. This is what results in quark 'confinement', the impossibility of producing free quarks. Because the QCD coupling constant has no $q=0$ limit, we have no natural point at which to settle on a conventional 'observed value'. We instead have to study the functional relation between q and the value of the coupling constant more carefully, and take this functional relationship into account in the whole discussion of renormalization. Thus the general functional relation between q and the coupling constant, known as the 'renormalization group', has become an important subject in contemporary quantum field theory.

4. RENORMALIZATION OF CHARGE AND A THIRD PROBLEM

As I indicated, an electron's charge, like its mass, must be renormalized. (This section will exclusively discuss the electron's charge. Similar things go for other 'coupling constants' which arise in quantum field theories.) We follow a procedure similar to the one we used for mass. We start with a 'bare' charge, e_0, which in this case is the charge an electron would have if it did not interact with itself.[1]

We then take interaction into account in a scheme of approximations. In this scheme, something which we can again think of as a self-interaction results in a divergent modification. We tell the same story of an 'infinite' bare charge balancing an 'infinite' self-interaction, yielding a finite result, which we call the renormalized charge, e_r, leaving behind a finite 'radiative correction' when $q \neq 0$. As before, we unpack talk about infinities by carrying out the discussion with a cut-off, L, which we take to infinity at the end. At each stage (for each value of L) we nail down the value of e_r as the observed value of the electron's charge.

I want to bring out some further aspects of renormalization by further considering charge renormalization. We may calculate the effects of the electromagnetic force in terms of an expression, M, which is again very useful in calculating observable quantities.[2]

We approximate M with a sequence of (we hope) more and more accurate estimates, $M_1, M_2, M_3, \ldots$. M_1 takes the form $-e_0^2 D$, the square of the bare charge times a term D, the details of which we can also ignore. So far, no divergent expressions occur.

However, the second-order correction introduces a divergent term, P_0, called an insertion term, so called because it gets inserted between two instances of the first-order term. The first and second approxi-

---

[1] Note for technicians: That is, if we omitted all vacuum polarization effects. Since e is the constant which couples the electron and photon fields, it makes no sense to talk, as we did for mass, about the value of the constant in the absence of all interaction.

[2] This time M is minus the charge squared times something called the 'photon propagator', but we are again best off not getting side-tracked on the question of just what sort of a quantity M is. The term, D, introduced next, is what the photon propagator would be in the absence of any interactions.

mations to M are:

$$M_1 = -e_0^2 D$$
$$M_2 = -e_0^2 D + (-e_0^2 D) P_0(-e_0^2 D).$$

The situation with higher-order corrections gets still more complicated. Part of the third-order corrections takes the form $(-e_0^2 D)P_0(-e_0^2 D)P_0(-e_0^2 D)$. But another part goes into modifying P_0. Still higher-order terms will modify P_0 further, and to bring out the pattern of how things proceed, let us just write P for the corrections that accrue to P_0 from all orders. (As I will comment below, this is mathematically not very satisfactory. But it fairly represents the kind of thinking in which almost all physicists indulge.) In sum, our third estimate, M_3, (which actually includes terms from all orders) takes the form

$$M_3 = -e_0^2 D + (-e_0^2 D) P(-e_0^2 D) + (-e_0^2 D) P(-e_0^2 D) P(-e_0^2 D),$$

and $M_4, M_5, \ldots$ follow the same pattern. In the limit, $M = (-e_0^2 D) \Sigma_{m=1}^{\infty} [P(-e_0^2 D)]^m$. Using the formula $1/(1+y) = 1 - y + y^2 - y^3 + \ldots$ with the value $y = P(-e_0^2 D)$ we get $M = \lim_{m \to \infty} (M_n) = (-e_0^2 D)/(1 + Pe_0^2 D)$. Next we rewrite P as $P = I/D + F$, where all the divergent parts of P are in I. This gives,

$$M = \frac{-e_0^2 D}{1 + e_0^2 I + Fe_0^2 D}. \tag{5.4}$$

The expression, F, contains all the finite radiative corrections. As with the parallel situation for mass, these terms are absent in situations like those in which we measure the charge. In other situations they provide a small correction. These corrections will not be relevant to the issues I want to discuss here, so we may streamline our presentation by just omitting the expression $Fe_0^2 D$, in effect by restricting consideration to situations in which the radiative corrections are 0.[3]

Dropping the term $Fe_0^2 D$ in (5.4), let us compare the expressions for the first approximation M_1 and the complete M:

$$M_1 = -e_0^2 D$$
$$M = \frac{-e_0^2 D}{1 + e_0^2 I}.$$

---

[3] One might be nervous about just ignoring the finite correction F, since it is multiplied by the bare charge, e_0^2, which is supposed to diverge. Readers should put this worry aside to get the drift of the ideas—and I will explain in the next footnote why they can legitimately do so.

Note that these differ only by the factor $Z = 1/(1 + e_0^2 I)$. Since we obtained M from M_1 by adding in all the higher-order corrections, we see that we can get the same effect by working only to first-order and, instead of using e_0^2, using

$$e_r^2 = Z e_0^2.$$

We call e_r the renormalized charge. As in the case of mass, we cannot turn off the self-interaction described by the higher-order terms. So we cannot observe e_0. The only charge we can observe is e_r, the one which carries with it all the higher-order self-interaction effects.[4]

As in the case of mass, the renormalization is infinite. I is divergent (but see note 6 below), this time comprised by a complex expression involving divergent integrals similar to those we encountered in the case of mass. However, we can tell the same sort of story we told for mass. We can think of e_0^2 and I as both diverging, but with the constant ratio $e_r^2 = Z e_0^2 = e_0^2/(1 + e_0^2 I)$. We can fill this out again by putting in cut-offs, working entirely with finite expressions, and finally letting the cut-off go to infinity after we obtain cut-off independent results.

How has the presentation of renormalization changed in going from the case of mass to the case of charge? No importance attaches to the use of differences in the case of mass vs. the use of ratios in the case of charge. The striking change lies in having treated only one order of the approximation at a time when renormalizing mass as opposed to having 'added' up all terms of the approximation in talking about charge. In fact, one can similarly add up all the terms for mass, but it is harder to illustrate because in so doing the problems of mass and charge renormalization get tangled up. However, in either case adding up all corrections raises some further questions.

We used the formula $1/(1+y) = 1 - y + y^2 - y^3 + \ldots$. But does the sequence converge? With y itself divergent, how could this possibly make any sense? The problem is not as bad as it seems. We learn the radius of convergence for this sequence to be $y < 1$. But the theory of analytic functions shows how to extend the area of convergence 'around' singularities (points at which the denominator becomes zero). When one puts in cut-offs, all this probably can be straightened

---

[4] The reason we do not have to worry about the e_0^2 in the term $F e_0^2 D$ in (5.4) is that this e_0^2 also gets converted to e_r^2 by renormalization. Starting with (5.4), multiply top and bottom by $Z = 1/(1 + e_0^2 I)$:

$$M = \frac{-Z e_0^2 D}{Z(1 + e_0^2 I + F e_0^2 D)} = \frac{-e_r^2 D}{1 + F e_r^2 D}.$$

out. I have never seen a discussion of the question. When things seem to work, physicists do not pause for such niceties.

The insertion term, P, presents a much more serious problem. It is a sequence which includes terms extracted from all orders of the basic approximation scheme. Physicists have succeeded in explicitly adding up the terms of this sequence only in special, very unphysical models. In the case of quantum electrodynamics and other physical theories, no one knows much about what P really looks like. The problem is not just that P's individual terms are divergent integrals. Even when the divergent integrals are made finite by putting in cut-offs, no one knows whether the sequence converges. In practice one uses at most the first few terms that enter into P.

The fact that P includes terms from all orders of the original approximation scheme raises a further question. We put $y = P(-e_0^2 D)$ into the sequence $1 - y + y^2 - y^3 + \ldots$. Since P itself includes terms from all orders, so doing plays very fast and lose with the order of summation, and order of summation can affect whether or not a sequence converges, and if it does converge, then to what limit. No one has hard-and-fast answers to these questions. Many practitioners seem confident that the original sequence of approximations at best satisfies a very weak sort of convergence called 'asymptotic convergence'. As far as I can tell much less still is known about what happens when the terms of the sequence are reorganized. Once again, when things are working well, there is just too much real physics to do for most practising physicists to take time for such questions.

The method I have presented for giving a closed expression for M in terms of P (or I) enables us to illustrate a further problem. While the sketch I am about to give over-simplifies the physics in a really extreme fashion, it may be useful in giving the non-technician some idea of the sorts of considerations which can arise. We make the simplifying assumption that I is independent of e_0, and that when we put a cut-off, L, on the divergent integrals, I diverges monotonically with L. These assumptions are satisfied if we include only the first contribution to I, in which case (writing $I(L)$ for I with the divergent integral cut off at L) $I(L) = C \ln(L)$, C a constant and ln the natural logarithm.

One can prove that $0 \leqslant Z \leqslant 1$.[5] Multiplying through by e_0^2 we get

$$0 \leqslant Z e_0^2 = e_r^2 = e_0^2/(1 + e_0^2 I) \leqslant e_0^2. \tag{5.5}$$

[5] I am going to present the following illustrative example on the presupposition that this is correct. The situation actually seems to be quite complex. See Schweber (1961), sects. 17b and 17c.

Three Problems of Renormalization

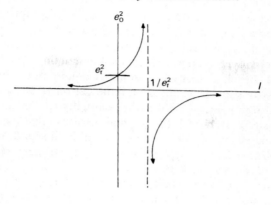

Fig. 5.1

and so

$$1/e_r^2 = 1/e_0^2 + I. \qquad (5.6)$$

The value of e_r^2 is fixed by observation. So e_0^2 is a function of I: As one can see from Fig. 5.1, (5.5) (using $0 \leqslant e_r^2 \leqslant e_0^2$) implies

$$0 \leqslant I < 1/e_r^2. \qquad (5.7)$$

The original theory told us that I, and the integrals of which it is comprised, were going to diverge.[6] But (5.7) says that this cannot happen!

What does this problem show? Strictly speaking, not very much because it is only known definitely to occur in very simple, unphysical models.[7]

In quantum electrodynamics and other physical theories, P is most unlikely to be independent of e_0^2, so nothing like the present simple argument will work. But some physicists suspect that this kind of problem, called a Landau singularity (Landau 1955; Landau et al. 1954), occurs in a quantum electrodynamics.

---

[6] As I mentioned, strictly speaking no one really knows what $P = I/D$ looks like. No one has added the sequence, even when its individual terms have been made finite by putting cut-offs on the divergent integrals. Moreover, since there has been a radical reorganization of the order of summation, the implications of a successful summing would still not be clear. But everyone seems to be feel very sure that, however this is straightened out, I will turn out to be divergent as $L \to \infty$. Thus reference here to 'what the theory tells us' involves a bit of a projection beyond what has been explicitly worked out.

[7] For example this situation arises in the Lee model. See Schweber (1961), pp. 365 ff.

As I mentioned, the fundamental theory plus approximation scheme asserts that $P = I/D$ diverges. So a Landau singularity constitutes an inconsistency. Something must be modified. The most natural alternative would be to look at a Landau singularity as telling us that the cut-off, L, really has an upper bound. According to this suggestion, we should not regard a finite cut-off as an intermediate step in the argument, where we will eventually let the cut-off go to infinity. Instead, we modify the original theory so that the problematic integrals get cut-off at some large but finite value of L. We will see shortly that we have independent and even stronger reasons for making a move very close to this.

A Landau singularity, as displayed in Fig. 5.1, suggests a further very interesting idea: There may be connections between the bounds of validity of the theory and the actual value of e_r. As one can easily read off from the graph, when a Landau singularity occurs the observed value of e_r^2 provides a bound on the value of I, and so a constraint on the problematic integrals, and so a constraint on the theory's domain of validity. That is the observed value of e_r^2 puts constraints on where the problematic integrals must be cut off. A kind of converse may intrigue us even more. The connection between the value of e_r^2 and the limits of validity of quantum electrodynamics might provide (parts of) an explanation for the value of e_r. For example when we have a Landau singularity, knowledge of the theory's domain of validity might set an upper bound on e_r. If some more accurate theory succeeds in giving us an account of just where and why quantum electrodynamics breaks down, this account, together with the connections between the value of e_r and the theory's domain of validity, might provide some kind of an explanation of the value of e_r.

Of course, real theories involve nothing remotely as simple as what I have described. But apparently theorists do toy with explanatory schemes of which the foregoing account provides a terribly oversimplified illustrative model.

5. INTERPRETING THE RENORMALIZATION STRATEGY

In texts, in classes, and in discussion with physicists, I have found three distinguishable attitudes towards the renormalization procedure. A few physicists, contrary to the spirit in which I have tried to present ideas here, seem to insist that we think of renormalization as, literally, discarding or absorbing actual infinities.[8] I call this the *Actual Infinities Approach* to renormalization.

The argument, I take it, is that when we cut off the integrals, thereby 'regularizing' the theory, we get a theory which we know is wrong. Nor can we evade the problem by citing that fact that there are other ways of regularizing the theory. These alternative regularization procedures serve the same function as cutting off the integrals, and those of these alternatives are technically much superior to the simple cut-off procedure which I have used in this simple exposition. But all known regularization procedures suffer in violating one or more basic physical principles, such as Lorentz invariance, gauge invariance, or unitarity (conservation of probability). The argument suggests that, since the regularized theories are known to be incorrect, we must reject them and boldly imbrace the idea of literally throwing away infinite terms.

This suggestion seems crazy, mathematically speaking. Yet this craziness does not give us a very conclusive reason for rejecting the idea. Physics has seen important precedents. Dirac seemed to be using crazy mathematics in employing his delta functions. But the method clearly proved useful, and mathematicians rose to the occasion by discovering distribution theory, in terms of which delta function practice makes sense. In fact we have much more substantial precedent in the calculus, which took centuries to grow secure foundational roots.

If we could not make sense of renormalization without new mathematics, I think we would have good reason to expect that the mathematics would eventually be forthcoming. But I am not very moved by the argument for thinking that we must reject alternative understandings of renormalization. Why should regularized theories satisfy all known theory constraints? All physical theories are approximations. If in an ample domain of application a theory no more than very slightly violates a condition which we believe nature exactly satisfies, that theory is not obviously cast into greater doubt than our other approximate theories.

I must qualify the last comment in one respect. If, in manipulating a theory, one wants to be free in appealing to constraints, one is well advised to use only constraints which the theory satisfies exactly.

---

[8] Though not clear-cut, one seems to discern this attitude in Feynman (1963), p. 145.

However, sophisticated forms of regularization meet this criterion. Indeed, search for sophisticated forms of regularization has been fuelled precisely by the need for a form which meets those constraints actually used in formal manipulations. But I do not see why a theory, known to be like all theories in being an approximation, needs exactly to meet unused constraints

I also have another reason for being sceptical of the argument for the actual infinities approach. The argument relies on the assumption that there is no fully satisfactory form of regularization, satisfactory in meeting all used and unused physical constraints. At present we do not know such a form. However, we shall see in a moment that we may have reason to believe than some such form none the less exists.

Most texts and classroom presentations avoid, as I have done here, the excess of actual infinities. In what I call the *Cut-offs Approach* one makes all expressions finite with one or another form of regularization, one absorbs the potentially divergent terms, and then one takes the limit. The procedure is mathematically sound but conceptually troubling. To smooth these troubles I have suggested thinking about what happens 'in the limit' as 'ideal points' to be understood in terms of the collection of all the real cases that occur along the way to the limit. No doubt, if helpful at all, these ideas will require much more careful development.

One may also ask whether the actual infinities and cut-off approaches are clearly distinct, or whether, when more is said, they might come to the same thing. To explain one way in which this might happen, I need to refer to a commonplace in field theory: The effects of self-interaction spread out in the space around a particle so that when experimenters probe exceedingly close to, say, an electron they avoid part of the self-interaction. Consequently, an electron shows a bigger charge from close up than it does from a long way away. (Probing close to a particle is equivalent to giving the impact parameter, q, a large value, which 'scales' the charge upward.) When I asked one actual infinities advocate what could be meant by an 'infinite' bare charge, he said: we know experimentally that when we probe closer and closer to an electron, we see a larger and larger charge. Now suppose that that should go on without any limit. I would now add that it makes no sense to probe to the point of sitting right on top of a point charge. So talk of an 'infinite' bare charge is starting to sound like talk about the larger and larger charge we might see as, without limit, we get closer and closer to the charge. And this now starts to

Three Problems of Renormalization

sound like the interpretation I want to suggest for the cut-offs approach.

I have not yet mentioned that I have found to be the most common attitude towards renormalization. The problem of infinite renormalization arises because of integrals which diverge as one integrates over a momentum variable from zero all the way to infinity. Now, integrating over all momenta means integrating over all energies. But we know that a theory like quantum electrodynamics must be wrong at very high energies. Such theories neglect the effects of, if nothing else, gravity, which is utterly negligible when we run elementary particles around in accelerators, but which invalidates the theory at exceedingly high energies. So we know that the infinite upper ends of those integrals, the parts which get 'cut off' in regularization, must be wrong anyway. So should we not just use a regularized version of the theory?

There is one problem with this suggestion: We do not know where to put in the cut-off. This is where we see a reason for letting the cut-off go to infinity at the end of the discussion. Assume that our theory approximates a correct, finite, theory—finite in the sense of being entirely free from divergent expressions. However, we do not at present know that theory, and we do not know where our theory, the theory we do have, leaves off being a good approximation. In what I call the *Mask of Ignorance Approach* to renormalization, we cover our ignorance by calculating only quantities which are independent of the exact value of the cut-off. We look only at quantities which are utterly insensitive to which large value of the cut-off one might chose to impose. We express this ignorance, and enforce the restriction to cut-off independent values, by letting the cut-off go to infinity at the end of the discussion. Taking the infinite cut-off limit comes to a way of saying that we do not know what large value the cut-off should have.

Because they neglect gravity and other interactions, we know that our theories must break down at very high energies. In the mind of a physicist, knowing that a theory breaks down always supports the faith that a better theory can be found. Surely it is reasonable to hope that such a better theory will be free of the divergence problems. If so, some of the better theory's finite expressions, those which correspond to the valid part of the presently divergent integrals, will constitute exactly that satisfactory regularization scheme which the advocate of actual infinities thinks will never be found. So faith in the mask of ignorance attitude gives one further reason (or at least motivation) for doubting the idea of discarding actual infinities as a permanent aspect of quantum field theory.

If the mask of ignorance approach should provide the soundest attitude towards infinite renormalization, would that mean that when we find a better theory, infinite renormalization must be quietly put aside as an embarrassing chapter in the history of physics? Most likely not. Even if use for infinite renormalization should disappear, it will have served as a sound and remarkably ingenious way of extracting information from an imperfect theory. And since all theories are imperfect, it is precisely such ingenuity which fuels much of the success of physics. Thus renormalization will always stand as an exemplar of good physics.

More importantly, renormalization may well remain as a prized component of the physicist's tool-kit. To begin with, even if the need for infinite renormalization should disappear, the need for finite renormalization will surely remain, for we have no reason to think that the effects of self-interaction on bare mass and bare coupling constants will ever be experimentally separable.[9]

So even in a finite theory renormalization will play an important role, complete with the phenomenon of running coupling constants and the renormalization group. More dramatically, even if, in principle, a finite theory is out there to be found, we might never find it. It might be too complicated. Or, if found, its complication might be so great as to make its use in calculating experimental results prohibitively hard, even impossible in practice. (Some seem to worry that a current candidate for a finite theory, superstring theory, might have such failings.) In such an eventually the requirement that a theory be renormalizable will remain as an essential condition on theories we can use[10], and infinite renormalization will always provide a central aspect of an important family of physical theories.

---

[9] This seems clear to the extent that the world is as described by quantum electrodynamics. In other theories the situation may be more complicated. For example in the more complex situation in quantum chromodynamics, where the running coupling constant plays an essential role, one might say that there is no longer such a clear distinction between the bare and 'dressed' coupling constant.

[10] I discuss the role of renormalizability as a theory constraint in more detail in 'Infinite Renormalization' referred to above.

REFERENCES

FEYNMAN, R. P. (1963): *Theory of Fundamental Processes* (New York: Benjamin Cummings).

LANDAU, L. D. (1955): 'On the Quantum Theory of Fields' in W. Pauli (ed.), *Niels Bohr and the Development of Physics* (London: Pergamon Press).

—— ABRIKOSOV, A. A., and KHALATNIKOV, I. M. (1954): 'The Removal of Infinities in Quantum Electrodynamics', *Doklady Akademii Nauk SSSR*, **95**, 497–9, 773–6, 1117–20; **96**, 261–263.

SCHWEBER, S. S. (1961): *An Introduction to Relativistic Quantum Field Theory* (New York: Harper & Row).

TELLER, P.: 'Infinite Renormalization', (Philosophy of Science forthcoming).

III
Covariance Principles in Quantum Field Theory

6
Hyperplane-dependent quantized Fields and Lorentz Invariance
GORDON N. FLEMING

1. SPACELIKE HYPERPLANES

The central concept that is used in a novel way in this investigation is that of a *space-like hyperplane*.[1] A spacelike hyperplane is a three-dimensional, metrically flat section of the flat Minkowski space-time continuum, such that any two points in the hyperplane are separated by a spacelike interval. For any such hyperplane, there is an inertial frame of reference in which all the points of the hyperplane are simultaneous, and all points simultaneous with any point of the hyperplane are in the hyperplane. The hyperplane is then said to be an *instantaneous hyperplane* for that inertial frame. However, it is the notion of a spacelike hyperplane as an *invariant geometrical construct*, with different descriptions from the standpoint of different intertial frames, that I wish to emphasize.

In terms of the Minkowski coordinates of any inertial frame, a spacelike hyperplane is composed of space-time points the coordinates of which satisfy a linear equation of the form (Fig. 6.1)

$$\eta^\mu x_\mu = \tau, \tag{6.1a}$$

where η^μ is a time-like unit four-vector satisfying

$$\eta^\mu \eta_\mu = 1 \tag{6.1b}$$

and, by convention, $\eta^0 \geq 1$. If (6.1a) is the equation for all the points in the hyperplane, then the hyperplane itself may be uniquely parameterized by the ordered pair $(\eta^\mu; \tau)$. If the hyperplane is instantaneous in the frame of reference being employed, then $\eta^\mu = (1, 0, 0, 0)$,

---

[1] Certainly the concept itself is not novel, being but a special case of the notion of a curvilinear space-like hypersurface introduced by S. Tomonaga (1946) and J. Schwinger (1948b) in the original manifestly covariant formulations of quantum field theory. See also Z. Bloch (1934).

© Gordon N. Fleming 1987.

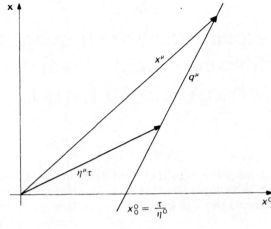

FIG. 6.1

and (6.1a) becomes $x_0 = \tau$. If the Minkowski coordinates of two inertial frames are related by the inhomogeneous Lorentz transformation,

$$x^{\mu'} = \Lambda^{\mu}_{\nu} x^{\nu} + a^{\mu} \tag{6.2a}$$

then the *same hyperplane*, which is parameterized by $(\eta^{\mu}; \tau)$ in the unprimed coordinate system, is parameterized by $(\eta^{\mu'}; \tau')$ in the primed coordinate system, where

$$\eta^{\mu'} = \Lambda^{\mu}_{\nu} \eta^{\nu}, \tag{6.2b}$$

and

$$\tau' = \tau + a_{\mu} \eta^{\mu'} = \tau + a_{\mu} \Lambda^{\mu}_{\nu} \eta^{\nu}. \tag{6.2c}$$

Since the four-vector $x^{\mu}_{(0)} \equiv \eta^{\mu} \tau$ satisfies (6.1a), it follows that $\eta^{\mu} \tau$ is the four-vector from the space-time origin of coordinates to the hyperplane, which points in the forward time-like direction orthogonal to the hyperplane. Any position four-vector, x^{μ}, for a point lying in the hyperplane, can be decomposed in the form (Fig. 6.1)

$$x^{\mu} = \eta^{\mu} \tau + q^{\mu}, \tag{6.3a}$$

where

$$\eta^{\mu} q_{\mu} = 0, \tag{6.3b}$$

and q^{μ} is parallel to the hyperplane. Under the inhomogeneous

Lorentz transformation (IHLT), (6.2a), we find

$$q^{\mu'} = \Lambda^\mu_\nu q^\nu + (a^\mu - \eta^{\mu'}\eta'_\nu a^\nu). \qquad (6.2d)$$

If two hyperplanes, $(\eta^\mu_1; \tau_1)$ and $(\eta^\mu_2; \tau_2)$, have $\eta^\mu_1 = \eta^\mu_2$ they will be called *parallel hyperplanes*, while if $\eta^\mu_1 \neq \eta^\mu_2$ they will be said to be *non-parallel* and to have different *orientations*. Suppose the point represented by x^μ lies in the $(\eta^\mu; \tau)$ hyperplane. They we can write (6.3a) with (6.3b) satisfied. Now consider the $(\tilde\eta^\mu; \tilde\tau)$ hyperplane with $\tilde\eta^\mu \neq \eta^\mu$. We have

$$\tilde\eta^\mu x_\mu = \tilde\eta^\mu \eta_\mu \tau + \tilde\eta^\mu q_\mu$$

and it is always possible to find q_μ such that (6.3b) holds and[2]

$$\tilde\eta^\mu \eta_\mu \tau + \tilde\eta^\mu q_\mu = \tilde\tau.$$

Thus, we have the algebraic demonstration of the geometrically intuitive result that non-parallel hyperplanes intersect.

If the equations (6.1) are written in three-dimensional notation, they can be combined to yield the result

$$\sqrt{(1+\boldsymbol\eta^2)}x^0 - \boldsymbol\eta\cdot\mathbf{x} = \tau. \qquad (6.4a)$$

This equation describes a moving two-dimensional plane, orthogonal to $\tilde{\boldsymbol\eta}$, which sweeps through the *space* of the inertial frame in question with superluminal velocity,

$$\mathbf{v} = \frac{\sqrt{(1+\boldsymbol\eta^2)}}{\boldsymbol\eta^2} c, \qquad (6.4b)$$

and crosses the spatial origin of coordinates at the time

$$t = \frac{\tau}{c\sqrt{(1+\boldsymbol\eta^2)}}. \qquad (6.4c)$$

The instantaneous hyperplanes for this frame correspond to $\boldsymbol\eta = 0$ or $\mathbf{v} = \infty$ at $t = \tau/c$.

2. HYPERPLANE DEPENDENCE IN QUANTUM FIELD THEORY

Hyperplane dependence of the dynamical variables of quantum theory, and, consequently, their eigenvectors, is the minimal general-

---

[2] From (6.3b) we have $q^0 = (\mathbf{q}\cdot\boldsymbol\eta)/\eta^0$. Substituting this into the equation we want to satisfy results in,

$$\left(\left[\frac{1+\tilde{\boldsymbol\eta}^2}{1+\boldsymbol\eta^2}\right]^{1/2}\boldsymbol\eta - \tilde{\boldsymbol\eta}\right)\cdot\mathbf{q} = \tilde\tau - (\tilde\eta\eta)\tau,$$

which, for $\tilde{\boldsymbol\eta} \neq \boldsymbol\eta$ clearly has a two-dimensional infinity of solutions for $\mathbf{q}$ and thus for q^μ.

ization of the concept of time dependence that is required to establish a manifestly Lorentz covariant formalism.[3] The reason for this is simply that the family of spacelike hyperplanes is the smallest set, which includes the instantaneous slices of the Minkowski continuum for any one inertial frame, and which is invariant, as a set, under the inhomogeneous Lorentz group (IHLG). The reason that hyperplane dependence has not previously become a prominent conceptual tool of theoretical physics is essentially twofold in nature. *First* is the widely held notion[4] that all global or spatially extended properties of single quantum mechanical particles can be expressed in terms of spatial integrals, over spatially local properties of such particles (such as the state function for detecting the particle at a particular point in space at a particular time), and such local properties, by virtue of reference to space-time points, require no reference to spacelike hyperplanes to be formulated in a Lorentz covariant manner. *Second*, one knows that contemporary fundamental theories of many-particle systems are expressed in terms of basic quantized fields that are themselves associated with simple points of space-time. Thus, once again the Lorentz covariance of the theory can be easily expressed in terms of the local fields without invoking hyperplane dependence explicitly. That global or spatially extended quantities would possess a form of hyperplane dependence is understood. But such hyperplane dependence is wholly secondary, derived, a consequence of definition in terms of basic local quantities. Such hyperplane dependence can be calculated from definitions if necessary and, fortunately, the necessity appears to be rare.

The *first* of these arguments is shown to be *erroneous* by a careful examination of the concept of spatial localization for relativisitc particles[5] and by the related demands of a Lorentz-invariant

---

[3] G. N. Fleming (1966).

[4] This 'widely held notion' is not sanctioned explicitly in the literature, so far as I know, since, in fact, as discussed below, it is erroneous; G. N. Fleming (1964). Nevertheless, I have encountered this notion in private discussions countless times and its principal origin appears to be the ease of overlooking the fact that, in elementary relativistic quantum theory, the effort to represent single-particle position probability densities by the time component of a local four-vector probability current density, in both the Klein–Gordon and Dirac formalisms, is fundamentally flawed. This point is overlooked, of course, because, in most texts, it is not made. See, however, S. S. Schweber (1961), ch. 2, and note 8 below.

[5] T. Newton and E. Wigner (1949); A. S. Wightman (1962).

formulation of the quantum theory of measurement.[6] I have contributed to the relevant analyses, in some detail, elsewhere.[7] Let me mention here three points. First, that the Minkowski coordinates of a Klein–Gordon or a Dirac wave function cannot be consistently interpreted as the eigenvalues of any self-adjoint position operator for the single particle, positive-energy state space.[8] Second, that, as Newton and Wigner showed many years ago,[9] an admissible position operator, uniquely defined in terms of its behaviour under the Euclidean symmetry group with reflections and self-commutation, displays counter-intuitive transformation properties under Lorentz transformations unless, as I have tried to show,[10] one recognizes the intrinsic hyperplane dependence of localization. Such recognition renders the otherwise counter-intuitive transformation properties expressible in manifestly covariant form devoid of all inconsistencies. Third, it should be mentioned that in the presence of spacelike separated measurement regions, state-vector assignment to those hyperplanes that do not intersect any such regions depends crucially on the way such hyperplanes order the measurement regions into an earlier set and a later set.[11] In this way, the apparent paradoxes of relativisitic quantum measurements can be eliminated.

The *second* argument against essential hyperplane dependence—the expressibility of all global or spatially extended quantities in terms of local fields enjoying local transformation properties—is not erroneous as far as it goes, but may be unnecessarily restrictive. The experience my students and I have gained,[12] in exploring the possibilities, allowed for interactions of particles with external potentials when hyperplane dependence is explicitly incorporated into the formalism, and suggests the possibility that consistent Lorentz-invariant quantum field theories with non-local interactions

---

[6] A representative sampling of contributions to this problem is provided by: I. Bloch (1967); K. Hellwig and K. Krause (1970); Y. Aharonov and D. Z. Albert (1980), (1981), (1984a), b); N. Giovannini (1983); M. L. G. Redhead (1983); S. Malin (1984); D. Dieks (1985); Shimony, A. (1985).
[7] G. N. Fleming (1964), (1966), (1985).
[8] This refers to note 4 above. If such an interpretation were possible, it would be possible to construct mutually orthogonal eigenfunctions for the Minkowski coordinates in the positive-energy state space. But such eigenfunctions do not exist. To build them one must superpose positive and negative energy states.
[9] See note 5 above.
[10] See (1964) and (1966) in note 7.
[11] See (1985) in note 7.
[12] G. N. Fleming (1986).

may be possible if the fields are hyperplane dependent. I will suggest below a model of such a field theory.

If we accept, for the moment, that such field theories can exist, what motivation is there for exploring them? What likelihood is there that they will be relevant to real physical systems?

The original motivation for exploring non-local interactions in quantum field theory, through the early 1950s[13], sprang from the concern that infinitely renormalized perturbation theory, the only systematic calculational technique for local quantum field theory at the time, would not be suitable for dealing with the strong interactions prevailing in nuclei. With time, a growing concern over the internal consistency of a theory requiring infinite renormalization and the long-standing recognition that local interactions generate that requirement, by producing poor convergence properties of the momentum space integrals of naïve perturbation theory, became the dominant motive for studying non-local field theories.[14] These efforts, however, largely floundered on the dual demands of the absence of causal anomalies and an unambiguous physical interpretation. Today infinite renormalization has been with us for a long time, and ever more sophisticated and elegant means of implementing it continue to be developed.[15] There is no question that at the level of comparing renormalized perturbation theory calculations with experiment, we are confronted with the most precise and accurate corroborations of theory to be found anywhere in science. The methods work wonderfully! Yet through all these years since Dyson, Feynmann, and Schwinger[16] formulated renormalization theory, it has never shed its fundamentally *ad hoc* character.[17] It remains a

[13] G. Wataghin (1934), (1935); McManus (1949); M. Fierz (1950); P. Kristensen and C. Møller (1952); A. Pais and G. F. Uhlenbeck (1950); D. R. Yennie, (1950); H. Yukawa (1950a, b); C. Bloch (1952); M. Chretien and R. E. Peierls (1953); W. Pauli (1953); E. C. G. Steuckelberg and G. Wanders (1954); J. Rayski (1954); T. Tati (1957); H. Yukawa (1965); D. A. Kirzhnits (1966); Y. Katayama, and H. Yukawa (1968). Attempts have also been made to interpret quantum electrodynamics as a non-local field theory in disguise: J. G. Valatin (1951); S. Mandelstam (1962); F. Rohrlich (1974). The formal structure of Valatin's study has much in common with the present paper.

[14] See note 13 above.

[15] G. Gallavotti (1985); J. Collins (1984).

[16] F. J. Dyson (1949); R. P. Feynman (1949); J. Schwinger (1948), (1949). For an excellent recent historical account of the period, see A. Pais (1986). For an illuminating account of one chapter of the period, see S. S. Schweber (1986).

[17] A rigorous axiomatization of renormalization theory was given by H. Epstein and D. Glaser (1973). This displays the preservation of the general principles of local quantum field theory through the renormalization process but, at the same time, obscures some of the simple algebraic properties of renormalized quantities.

recipe for extracting finite results from an infinity-plagued formalism by cancelling the infinities against one another systematically. What is wanted is a formulation of non-trivial interacting quantum field theory that never encounters the infinities in the first place. At various times the achievements in this direction of the constructive quantum field theorists have been impressive.[18] But they have never made much progress with non-trivial interactions in $3+1$ space-time dimensions; and the apparent specificity of their methods to the space-time dimensions of their models lends an air of unreality to the endeavour from the standpoint of the physicist. For the sake of simplicity, I, too, will present a model field theory in reduced dimensions, but the ideas and methods for constructing the model generalize trivially to $3+1$ dimensions.

The recent investigations of the quantum field theories of supersymmetric strings in higher dimensional space-times promise to shed light on our problem since some models of such theories have been shown to be finite up to second order in the loop expansion version of perturbation theory.[19] Still, two-loop finiteness is a long way from the property of an exact solution.[20] Consequently, there may still be merit in the comparatively mundane examination of non-local field theory of the hyperplane-dependent variety.

As to the physical relevance of such models for real physical systems, there are two easily identified possibilities. At the present level of precision in the experimental testing of specific local quantum field theories, one cannot rule out the possibility that non-local interactions with an order parameter of the Planck length or somewhat larger may be at the bottom of things.[21] Such interactions could yield a finite theory, the long-distance consequences of which we extract from a local model via renormalization theory. The second possibility is that, even if local field theory is basically correct, and even if we must accept infinite renormalization as an essential aspect of a fundamental theory, we may still find that a model with non-local interactions can simulate non-perturbative features of such a funda-

---

[18] J. Ehlers (1973); E. Seiler (1982).
[19] J. H. Schwarz (1982); L. Clavelli and A. Halprin (1986).
[20] S. Glashow and P. Ginsparg (1986).
[21] The conceptual distinction between non-local interactions of fields and field-like theories of interactions between non-local objects such as strings, is not as sharp as this sentence would suggest. In any case, experimental testing of quantum field theories, local or otherwise, is currently regarded as having progressed only to spatial dimensions of the order of 10^{-16} cm.

mental theory in a more tractable way. In either of these situations, the first step must be to demonstrate the possibility of an internally consistent, physically interpretable, Lorentz-invariant quantum field theory with non-local interactions. This paper is offered as a step in that direction.

3. LORENTZ-INVARIANT QUANTUM THEORIES AND LOCAL FIELDS

In formulating a quantum theory invariant under the IHLG, it is mandatory to secure the existence of unitary operators,

$$\hat{U}(\Lambda, a),$$

which represents the IHLTs in the state space of the system of interest.[22] In other words, if a transformation between inertial frames of reference has the form,

$$x^{\mu'} = \Lambda^\mu_\nu x^\nu + a^\mu \tag{6.5}$$

where

$$g_{\mu_1\mu_2}\Lambda^{\mu_1}_{\nu_1}\Lambda^{\mu_2}_{\nu_2} = g_{\nu_1\nu_2}, \tag{6.6}$$

then the corresponding transformation of state vectors is

$$|\Psi'\rangle = \hat{U}(\Lambda, a)|\Psi\rangle. \tag{6.7}$$

It follows from this that the $\hat{U}$'s must satisfy[23]

$$\hat{U}(\Lambda_2, a_2)\hat{U}(\Lambda_1, a_1) = \hat{U}(\Lambda_2\Lambda_1, \Lambda_2 a_1 + a_2), \tag{6.8}$$

i.e. they form a unitary representation of the IHLG. It then follows from Lie group theory that the $\hat{U}$s have the form[24]

$$\hat{U}(\Lambda, a) = e^{\frac{i}{\hbar}\hat{P}^\mu a_\mu} e^{-\frac{i}{2\hbar}\hat{M}^{\mu\nu}\omega_{\mu\nu}(\Lambda)}. \tag{6.9}$$

The $\omega_{\mu\nu}(\Lambda)$ are determined by the requirements that

$$\omega_{\mu\nu}(\Lambda) = -\omega_{\nu\mu}(\Lambda) \tag{6.10a}$$

and

$$\omega_{\mu\nu}(\Lambda^n) = n\omega_{\mu\nu}(\Lambda) \tag{6.10b}$$

for any Λ, while the self-adjoint operators $\hat{P}^\mu$ and $\hat{M}^{\mu\nu}$ must satisfy

$$[\hat{P}^\mu, \hat{P}^\nu] = 0, \quad [\hat{M}^{\mu\nu}, \hat{P}^\lambda] = i\hbar(g^{\nu\lambda}\hat{P}^\mu - g^{\mu\lambda}\hat{P}^\nu), \tag{6.11a}$$

and

$$[\hat{M}^{\mu\nu}, \hat{M}^{\lambda\rho}] = i\hbar(g^{\nu\lambda}\hat{M}^{\mu\rho} - g^{\nu\rho}\hat{M}^{\mu\lambda} + g^{\mu\rho}\hat{M}^{\nu\lambda} - g^{\mu\lambda}\hat{M}^{\nu\rho}). \tag{6.11b}$$

[22] P. A. M. Dirac (1962).
[23] A phase factor could appear on the right-hand side. These phase factors can always be chosen to be ± 1. See E. P. Wigner (1939).
[24] S. Wightman (1960).

Thus, in order to secure the existence of the $\hat{U}$s, it is necessary and sufficient to be able to build the generators of the IHLG, the $\hat{M}^{\mu\nu}$ and the $\hat{P}^\mu$.

If $\hat{\phi}_\alpha(x)$ denotes the components (tensorial, spinorial, whatever) of a multi-component local field operator associated with the Minkowski coordinates x^μ, and if the $\hat{\phi}_\alpha(x)$ transform under the IHLG according to

$$(\Psi'|\hat{\phi}_\alpha(x')|\Psi') = S_{\alpha\beta}(\Lambda)(\Psi|\hat{\phi}_\beta(x)|\Psi), \tag{6.12}$$

where x' and $|\Psi'\rangle$ are given by (6.5) and (6.7), then one can show that the $\hat{\phi}_\alpha(x)$ must satisfy the commutation relations[25]

$$[\hat{\phi}_\alpha(x), \hat{P}_\mu] = i\hbar \frac{\partial}{\partial x^\mu} \phi_\alpha(x), \tag{6.13a}$$

$$[\hat{\phi}_\alpha(x), \hat{M}_{\mu\nu}] = i\hbar \left\{ \left(x_\mu \frac{\partial}{\partial x^\nu} - x_\nu \frac{\partial}{\partial x^\mu} \right) \hat{\phi}_\alpha(x) + \sum_{\mu\nu,\alpha\beta} \hat{\phi}_\beta(x) \right\}, \tag{6.13b}$$

where the numbers, $\Sigma_{\mu\nu,\alpha\beta}$, are found in $S_{\alpha\beta}(\Lambda)$ in the form

$$S_{\alpha\beta}(\Lambda) = (e^{\frac{1}{2}\Sigma_{\mu\nu}\omega^{\mu\nu}(\Lambda)})_{\alpha\beta}. \tag{6.14}$$

If the $\hat{\phi}_\alpha(x)$ exhaust the basic dynamical variables of the theory, the fundamental construction that must be effected is the expression of the generators $\hat{P}_\mu$ and $\hat{M}_{\mu\nu}$ in terms of the $\hat{\phi}_\alpha(x)$, so that all the commutation relations we have considered are satisfied.

To carry out this construction requires that the equal-time commutation relations among the basic fields and their derivatives be specified. The commutation relations are specified at equal times, since the initial conditions that determine a unique solution of the interacting field equations consist in the (operator) values of the fields and their derivatives at a definite time. Of course, such specification is frame dependent since instantaneous slices of the Minkowski continuum, i.e. definite times, are frame dependent. One can avoid the frame-dependent language by recognizing that frame-dependent specification of fields, their derivatives, and their commutators at definite times is equivalent to specification of fields, their derivatives, and their commutators on definite spacelike hyperplanes of any orientation. Thus, already in local quantum field theory, a germ of hyperplane dependence is present.[26] It is carried by the basic

---

[25] R. Streater and A. S. Wightman (1964); see also note 24.

[26] Though not couched in exactly these terms, statements to this effect can be found in J. M. Jauch and F. Rohrlich (1955), 9 and 23 and J. D. Bjorken and S. D. Drell (1965), 73 and 79, as well as other textbooks.

commutation relations which, for bosonic fields free of constraints, take the form

$$\delta(\eta x' - \eta x)[\hat{\phi}_\alpha(x'), \hat{\phi}_\beta(x)] = \delta(\eta x' - \eta x)\left[\frac{\partial}{\partial x^{\mu'}}\hat{\phi}_\alpha(x'), \frac{\partial}{\partial x^\nu}\hat{\phi}_\beta(x)\right] = 0,$$

$$\delta(\eta x' - \eta x)[\hat{\phi}_\alpha(x'), \frac{\partial}{\partial x^\mu}\hat{\phi}_\beta(x)] = i\hbar\,\delta_{\alpha\beta}\eta_\mu\delta^4(x' - x), \qquad (6.15)$$

for any time-like unit four-vector.

The simplest non-trivial example of the construction of the generators $\hat{P}_\mu$ and $\hat{M}_{\mu\nu}$ for a field theoretic system is provided by the system of a self-interacting neutral scalar field with non-derivative coupling. In this case we have[27]

$$\hat{P}^\mu = \int d^4x\,\delta(\eta x - \tau)\,\hat{\theta}^{\mu\nu}(x)\,\eta_\nu, \qquad (6.16a)$$

$$\hat{M}^{\mu\nu} = \int d^4x\,\delta(\eta x - \tau)\{x^\mu\hat{\theta}^{\nu\lambda}(x) - x^\nu\hat{\theta}^{\mu\lambda}(x)\}\,\eta_\lambda, \qquad (6.16b)$$

where

$$\hat{\theta}^{\mu\nu}(x) \equiv \hat{\theta}^{\mu\nu}_{(0)}(x) + g^{\mu\nu}\hat{v}(x), \qquad (6.17)$$

is the stress-energy-momentum tensor for the system, with

$$\hat{\theta}^{\mu\nu}_{(0)}(x) = \partial^\mu\hat{\phi}(x)\,\partial^\nu\hat{\phi}(x) - \frac{g^{\mu\nu}}{2}(\partial^\rho\hat{\phi}(x)\,\partial_\rho\hat{\phi}(x) - K^2\hat{\phi}(x)^2), \qquad (6.18)$$

and

$$\hat{v}(x) = f(\hat{\phi}(x)). \qquad (6.19)$$

The equation of motion for the field in this case is

$$(\Box + K^2)\hat{\phi}(x) = -\frac{\partial f(\hat{\phi}(x))}{\partial \hat{\phi}(x)}. \qquad (6.20)$$

From this field equation, it follows that the stress-energy-momentum tensor is divergenceless

$$\partial_\mu\hat{\phi}^{\mu\nu}(x) = 0, \qquad (6.21)$$

and from this in turn, it follows that $\hat{P}^\mu$ and $\hat{M}^{\mu\nu}$ do not depend on the

---

[27] In fact, it is always possible to write the generators, P^μ and $M^{\mu\nu}$, in this form in any quantum field theory derivable from the specification of a Lagrangian. See (1955) in note 26. For fields other than self-interacting neutral scalar, of course, the expressions (6.17–19) for the stress-energy-momentum tensor are inapplicable.

hyperplane on which they are evaluated, i.e. they are conserved

$$\frac{\partial}{\partial \tau} \hat{P}^\mu = \frac{\partial}{\partial \tau} \hat{M}^{\mu\nu} = 0; \quad \frac{\partial}{\partial \eta^\lambda} \hat{P}^\mu = \frac{\partial}{\partial \eta^\lambda} \hat{M}^{\mu\nu} = 0. \tag{6.22}$$

4. HYPERPLANE-DEPENDENT QUANTIZED FIELDS

Of the equations in the preceding section, the first one to be altered by hyperplane dependence of the fields is (6.12), which now becomes[28]

$$(\Psi'|\hat{\phi}_\alpha(x',\eta')|\Psi') = S_{\alpha\beta}(\Lambda)(\Psi|\hat{\phi}_\beta(x,\eta)|\Psi). \tag{6.23}$$

The notation, $\hat{\phi}_\alpha(x,\eta)$, signifies that it is not sufficient to specify the point of Minkowski space-time to determine the field; one must also specify the orientation, η^μ, of a particular spacelike hyperplane passing through the point x^μ. The fields are not defined at points of space-time, but at *points on hyperplanes* in space-time. In this context, two hyperplane points (x, η) and (x, η'), having the same Minkowski coordinates, x, are not to be regarded as physically the same point. Technically, one can say that the fields are defined on a seven-dimensional homogeneous space of the IHLG.[29]

From (6.23), a consideration of infinitesimal transformations yields the commutation relations

$$[\hat{\phi}_\alpha(x,\eta), \hat{P}_\mu] = i\hbar \frac{\partial}{\partial x^\mu} \hat{\phi}_\alpha(x,\eta) \tag{6.24a}$$

$$[\hat{\phi}_\alpha(x,\eta), \hat{M}_{\mu\nu}] = i\hbar \left(x_\mu \frac{\partial}{\partial x^\nu} - x_\nu \frac{\partial}{\partial x^\mu} + \eta_\mu \frac{\partial}{\partial \eta^\nu} - \eta_\nu \frac{\partial}{\partial \eta^\mu} \right) \hat{\phi}_\alpha(x,\eta)$$

$$+ i\hbar \sum_{\mu\nu,\alpha\beta} \hat{\phi}_\beta(x,\eta), \tag{6.24b}$$

where, as before, (6.14) holds. The basic field–field commutation relations, in the absence of derivative coupling, will be assumed to

---

[28] See note 3 above.
[29] G. N. Fleming (1970). This is just one example of a type of structure that is typical of field theories defined on homogeneous spaces of the IHLG. Study of such field theories briefly became a minor industry in the late 1960s as an early version of the search for unified field theories of the many particle types then known. See F. Lurcat (1964); H. C. Corben (1965); T. Takabayashi (1965), (1966); Y. Nambu (1966), (1967); C. Fronsdal (1968); G. Feldman and P. T. Matthews (1967); A. O. Barut and H. Kleinent (1967); H. Bebie and H. Leutwyler (1967); S. J. Chang and L. O'Raifeataish (1968); E. Abers, I. T. Grodsky and R. E. Norton (1967); I. T. Grodsky and R. Streater (1968); F. Ardalan and G. N. Fleming (1975); C. Boyer and G. N. Fleming (1974). The earliest study of fields of this type appears to be that of E. Majorana (1932).

retain their previous form,

$$\delta(\eta x' - \eta x)[\hat{\phi}_\alpha(x',\eta), \hat{\phi}_\beta(x,\eta)] = 0, \tag{6.25a}$$

$$\delta(\eta x' - \eta x)\left[\frac{\partial}{\partial x^{\mu'}}\hat{\phi}_\alpha(x',\eta), \frac{\partial}{\partial x^\nu}\hat{\phi}_\beta(x,\eta)\right] = 0, \tag{6.25b}$$

$$\delta(\eta x' - \eta x)\left[\hat{\phi}_\alpha(x',\eta), \frac{\partial}{\partial x^\nu}\hat{\phi}_\beta(x,\eta)\right] = i\hbar\,\delta_{\alpha\beta}\eta_\nu\delta^4(x'-x), \tag{6.25c}$$

although these equal hyperplane commutation relations, by virtue of the η dependence carried by the fields, are not as restrictive as in the case of local fields.

Before trying to build generators $\hat{P}_\mu$ and $\hat{M}_{\mu\nu}$, which would consistently force non-trivial η-dependence on the fields, a few comments on the interpretation of the particle aspect of such fields is in order. For a neutral scalar field, $\hat{\phi}(x,\eta)$, we introduce the creation and annihilation operators, $\hat{A}^+(k;\eta,\tau)$ and $\hat{A}(k;\eta,\tau)$ in the expansions

$$\hat{\phi}(x,\eta) \equiv (2\pi\hbar)^{-3/2}\hbar\int\frac{d^4k}{\omega(k)}\,\delta(\eta k)\{\hat{A}(k;\eta,\eta x)e^{\frac{i}{\hbar}k\cdot x} + \hat{A}^+(k;\eta,\eta x)e^{-\frac{i}{\hbar}k\cdot x}\} \tag{6.26a}$$

$$\eta\frac{\partial}{\partial x}\hat{\phi}(x,\eta) \equiv (2\pi\hbar)^{-3/2}i\int d^4k\,\delta(\eta k)\{\hat{A}(k;\eta,\eta x)e^{\frac{i}{\hbar}k\cdot x} - A^+(k;\eta,\eta x)e^{-\frac{i}{\hbar}k\cdot x}\}, \tag{6.26b}$$

where $\omega(k) \equiv \sqrt{(k^2 + K^2)}$. As a consequence of the basic commutation relations, it follows that

$$\delta(\eta k' - \eta k)[\hat{A}(k';\eta,\tau), \hat{A}(k;\eta,\tau)] = 0, \tag{6.27a}$$

$$\delta(\eta k' - \eta k)[\hat{A}(k';\eta,\tau), \hat{A}^+(k;\eta,\tau)] = \omega(k)\delta^4(k'-k), \tag{6.27b}$$

and

$$[\hat{A}(k;\eta,\tau), \hat{P}_\mu] = (k_\mu - \eta_\mu\eta k)\hat{A}(k;\eta,\tau) + i\hbar\frac{\partial}{\partial\tau}\hat{A}(k;\eta,\tau)\eta_\mu, \tag{6.27c}$$

$$[\hat{A}(k;\eta,\tau), \hat{M}_{\mu\nu}] = i\hbar\left(k_\mu\frac{\partial}{\partial k^\nu} - k_\nu\frac{\partial}{\partial k^\mu} + \eta_\mu\frac{\partial}{\partial\eta^\nu} - \eta_\nu\frac{\partial}{\partial\eta^\mu}\right)\hat{A}(k;\eta,\tau). \tag{6.27d}$$

These commutation relations secure the interpretation of the $\hat{A}^+$ and $\hat{A}$s as creation and annihilation operators for a quanta on the (η,τ) hyperplane, which carries hyperplane momentum $\hat{P}_\mu - \eta_\mu\eta\hat{P}$ in the amount $k_\mu - \eta_\mu\eta k$. One must be very cautious, however, about identifying these quanta with traditional particles of definite spin and mass. Even if one introduces interactions in such a way as to guarantee a local scalar evolution of the field as the copling constant goes to zero—as will be the case in the example of the next section—the asymptotic behaviour of the field in the distant past and future, with the interaction kept on throughout, can, in principle, correspond to the free evolution of a *family* of local fields of different spins and spin-dependent masses. This is achieved through the expansion[30]

$$\tilde{\phi}(p,\eta) = \sum_{\substack{s=0 \\ n}}^{\infty} h_{\mu 1} \ldots h_{\mu s} \rho_{s,n}(\eta p) \tilde{\phi}_n^{\mu_1 \cdots \mu_s}(p), \qquad (6.28)$$

where

$$\tilde{\phi}(p,\eta) = (2\pi\hbar)^{-2} \int d^4x \hat{\phi}(x,\eta) e^{ipx}, \qquad (6.29)$$

and

$$h_\mu \equiv \eta_\mu - p_\mu \frac{\eta p}{p^2}. \qquad (6.30)$$

The $\rho_{s,n}(\eta p)$ are a complete set of appropriately normalized functions determined from the equation of motion for $\hat{\phi}$, while the fields, $\tilde{\phi}_n^{\mu_1 \cdots \mu_s}(p)$, are completely symmetric, traceless, transverse (in the sense that $p_{\mu_1} \tilde{\phi}_n^{\mu_1 \cdots \mu_s}(p) = 0$) fields describing particles of spin s. The asymptotic motion of the original hyperplane-dependent field $\hat{\phi}(x,\eta)$, is governed by an equation of the form[31] (in those cases where this analysis would apply)

$$f(\square, \Omega^2) \hat{\phi}(x,\eta) = 0, \qquad (6.31)$$

where

$$\Omega_\mu \equiv \varepsilon_{\mu\alpha\beta\gamma} \eta^\alpha \frac{\partial}{\partial \eta_\beta} \frac{\partial}{\partial x_\gamma}, \qquad (6.32)$$

and Ω^2 corresponds to the Pauli–Lubyanski Casimir invariant of the IHLG in the same way that the d'Alembertian, $\square$, corresponds to $\hat{P}^2$. The mass spectra of the particles is determined by the equation

$$f\left(-\frac{p^2}{\hbar^2}, s(s+1)\frac{p^2}{\hbar^2}\right) \hat{\phi}_n^{\mu_1 \cdots \mu_s}(p) = 0. \qquad (6.33)$$

We are finally ready to consider a specific model with non-trivial hyperplane dependence enforced by the presence of non-local

---

[30] See note 29 above.
[31] See note 26 above.

interactions. Since the most difficult task in building such a model is that of insuring internal consistency among all the relations of the model we will simplify our task by considering a model in $1+1$ dimensional space-time. The generalization of the model to $3+1$ dimensional space-time is, I believe, straightforward, but I have not yet carried out that generalization.

5. NON-LOCAL INTERACTIONS IN $1+1$ DIMENSIONAL SPACE-TIME

In $1+1$ dimensional space-time, the time-like unit two-vector, η^μ, is conveniently expressed as[32]

$$\eta^\mu = (\eta^0, \eta^1) = (\cosh\theta, \sinh\theta). \tag{6.34}$$

Infinitesimal reorientations of hyperplanes (they are now simply lines but we will continue to call them hyperplanes), represented by $\delta\eta^\mu$ become

$$\delta\eta^\mu = (\sinh\theta, \cosh\theta)\delta\theta, \tag{6.35}$$

and, since we must have

$$\delta\eta^\mu \frac{\partial}{\partial \eta^\mu} = \delta\theta \frac{\partial}{\partial \theta} \tag{6.36a}$$

and

$$\eta^\mu \frac{\partial}{\partial \eta^\mu} = 0, \tag{6.36b}$$

it follows that

$$\frac{\partial}{\partial \eta^\mu} = (-\sinh\theta, \cosh\theta) \frac{\partial}{\partial \theta}. \tag{6.37}$$

For some purposes it is useful to employ the fundamental antisymmetric tensor density

$$\varepsilon^{\mu\nu} = g_1^\mu g_0^\nu - g_1^\nu g_0^\mu, \tag{6.38}$$

to write

$$\delta\eta^\mu = \varepsilon^{\mu\nu}\eta_\nu \delta\theta, \tag{6.39a}$$

and

$$\frac{\partial}{\partial \eta^\mu} = -\varepsilon_{\mu\nu}\eta^\nu \frac{\partial}{\partial \theta}. \tag{6.39b}$$

---

[32] See (1985), Lect. 4 in note 7 above.

For any two-vector V^μ, we can write
$$V^\mu \equiv \eta^\mu V + \varepsilon^{\mu\nu}\eta_\nu W, \tag{6.40a}$$
where
$$V = V^\mu \eta_\mu; \quad W = -\varepsilon_{\mu\nu} V^\mu \eta^\nu. \tag{6.40b}$$

If this analysis is performed on the IHLG generators $\hat{P}^\mu$, we have
$$\hat{P}^\mu \equiv \eta^\mu \hat{H} + \varepsilon^{\mu\nu}\eta_\nu \hat{K}, \tag{6.41}$$
where $\hat{H}$ generates infinitesimal translations normal to the hyperplane, and $\hat{K}$ generates infinitesimal translations along the hyperplane. The antisymmetric homogeneous generator $\hat{M}_{\mu\nu}$ can be written as
$$\hat{M}_{\mu\nu} \equiv \varepsilon_{\mu\nu} \hat{N}, \tag{6.42}$$
and the commutation relations between these generators are
$$[\hat{H}, \hat{K}] = 0,$$
$$[\hat{N}, \hat{K}] = i\hbar \hat{H}, \quad [\hat{N}, \hat{H}] = i\hbar \hat{K}. \tag{6.43}$$

Since the generators, $\hat{P}^\mu$ and $\hat{M}^{\mu\nu}$, are hyperplane-independent, it follows that $\hat{N}$ is also, while $\hat{H}$ and $\hat{K}$ can be shown to satisfy
$$\frac{\partial \hat{H}}{\partial \theta} = -\hat{K}; \quad \frac{\partial \hat{K}}{\partial \theta} = -\hat{H}. \tag{6.44}$$

In other words
$$\hat{H}(\theta) = \cosh\theta\, \hat{H}(0) - \sinh\theta\, \hat{K}(0),$$
$$\hat{K}(\theta) = \cosh\theta\, \hat{K}(0) - \sinh\theta\, \hat{H}(0), \tag{6.45}$$
where
$$\hat{H}(0) = \hat{P}^0, \quad \hat{K}(0) = \hat{P}^1.$$

For the hyperplane-dependent field, $\hat{\phi}(x, \eta)$, it will be convenient to introduce the decomposition
$$x^\mu \equiv \eta^\mu \tau + \varepsilon^{\mu\nu}\eta_\nu r, \tag{6.46}$$
and write
$$\hat{\phi}(x, \eta) \equiv \hat{\phi}(r; \theta, \tau). \tag{6.47}$$
Under the IHLT
$$x^{\mu'} = \Lambda^\mu_\nu x^\nu + a^\mu, \tag{6.48}$$
with
$$\Lambda^0_0 = \Lambda^1_1 = \cosh\alpha, \quad \Lambda^0_1 = \Lambda^1_0 = \sinh\alpha, \tag{6.49}$$

we find
$$\theta' = \theta + \alpha, \tag{6.50a}$$
$$\tau' = \tau + (\cosh \theta') a^0 - (\sinh \theta') a^1, \tag{6.50b}$$
$$r' = r + (\cosh \theta') a^1 - (\sinh \theta') a^0. \tag{6.50c}$$

It follows from this and the scalar field version of (6.24) that the Heisenberg equations of motion for the field are

$$-i\hbar \frac{\partial}{\partial r} \hat{\phi}(r;\theta,\tau) = [\hat{\phi}(r;\theta,\tau), \hat{K}(\theta)], \tag{6.51a}$$

$$i\hbar \frac{\partial}{\partial \tau} \hat{\phi}(r;\theta,\tau) = [\hat{\phi}(r;\theta,\tau), \hat{H}(\theta)], \tag{6.51b}$$

and

$$i\hbar \frac{\partial}{\partial \theta} \hat{\phi}(r;\theta,\tau) = [\hat{\phi}(r;\theta,\tau), \hat{N}]. \tag{6.51c}$$

We must satisfy these commutation relations and those holding between the generators in our construction of the model.

To begin the construction, we introduce the canonical conjugate field $\hat{\pi}(r;\theta,\tau)$, which must satisfy, along with $\hat{\phi}(r;\theta,\tau)$, the equal hyperplane commutation relations

$$[\hat{\phi}(r';\theta,\tau), \hat{\phi}(r;\theta,\tau)] = [\hat{\pi}(r';\theta,\tau), \hat{\pi}(r;\theta,\tau)] = 0 \tag{6.52a}$$
$$[\hat{\phi}(r';\theta,\tau), \hat{\pi}(r;\theta,\tau)] = i\hbar \delta(r'-r). \tag{6.52b}$$

As in local field theory, the relationship between $\hat{\pi}$ and the derivatives of $\hat{\phi}$ is determined by the form of the generators when expressed in terms of $\hat{\phi}$ and $\hat{\pi}$. For the non-derivative coupling that we will assume, we will obtain the standard result that

$$\hat{\pi}(r;\theta,\tau) = \frac{\partial}{\partial \tau} \hat{\phi}(r;\theta,\tau). \tag{6.53}$$

To ensure that the hyperplane dependence of our model stems solely from non-local interactions, we will write the generators in the form,

$$\hat{K} = \hat{K}_{(0)}, \tag{6.54a}$$
$$\hat{H} = \hat{H}_{(0)} + \hat{V}, \hat{N} = \hat{N}_{(0)} + \hat{U}, \tag{6.54b}$$

where $\hat{V}$ and $\hat{U}$ are proportional to the interaction coupling strength, and $\hat{K}_{(0)}, \hat{H}_{(0)}$, and $\hat{N}_{(0)}$ take the standard form for non-interacting, local, scalar fields:[33]

$$\hat{K}_{(0)} = -\int dr\, \hat{\pi}\, \frac{\partial}{\partial r}\, \hat{\phi}, \tag{6.55a}$$

$$H_{(0)} = \frac{1}{2}\int dr\, \left\{\hat{\pi}^2 + \left(\frac{\partial}{\partial r}\, \hat{\phi}\right)^2 + \kappa^2 \hat{\phi}^2\right\}, \tag{6.55b}$$

$$N_{(0)} = \frac{1}{2}\int dr\, r\, \left\{\hat{\pi}^2 + \left(\frac{\partial}{\partial r}\, \hat{\phi}\right)^2 + \kappa^2 \hat{\phi}^2\right\} - \tau \hat{K}_{(0)}. \tag{6.55c}$$

If $\hat{V} = \hat{U} = 0$, these generators yield (6.53), and

$$(\Box + \kappa^2)\, \hat{\phi} = 0, \tag{6.56a}$$

$$\frac{\partial}{\partial \theta}\, \hat{\phi} = \left(r\, \frac{\partial}{\partial \tau} + \tau\, \frac{\partial}{\partial r}\right) \hat{\phi}, \tag{6.56b}$$

and

$$\frac{\partial}{\partial \theta}\, \hat{\pi} = \left(r\, \frac{\partial}{\partial \tau} + \tau\, \frac{\partial}{\partial r}\right) \hat{\pi} + \frac{\partial}{\partial r}\, \hat{\phi}. \tag{6.56c}$$

The last two equations express, in terms of the hyperplane variables $(r; \theta, \tau)$, the local nature of $\hat{\phi}$, (6.56b), and the deviation from locality of $\hat{\pi}$, (6.56c), that is demanded by (6.53).

We now turn on the interaction in the form[34]

$$V(\theta, \tau) = G \int dr\, \left(\int dr'\, f(r - r')\, \hat{\phi}(r'; \theta, \tau)\right)^4, \tag{6.57a}$$

and

$$U(\theta, \tau) = G \int dr\, r\, \left(\int dr'\, f(r - r')\, \hat{\phi}(r'; \theta, \tau)\right)^4 + \hat{\Delta}(\theta, \tau). \tag{6.57b}$$

If $f(r - r') = \delta(r - r')$, we would have standard local $\hat{\phi}^4$ self-interaction, and we know that consistency would hold with $\hat{\Delta} = 0$. Upon allowing f to deviate from a delta function, we will see that consistency of the formalism demands $\hat{\Delta} \neq 0$, and completely determines $\frac{\partial}{\partial \tau}\, \Delta$.

The commutators of $\hat{\phi}$ and $\hat{\pi}$ with $\hat{H}$, taken together, yield (6.53)

---

[33] Although we will not focus on them here, the usual concerns about normal ordering in order to render these equations unambiguous, apply.

[34] See note 33.

and the equation of motion for $\hat{\phi}$,

$$(\Box + \kappa^2)\hat{\phi}(r;\theta,\tau) = -4G\int dr' f(r'-r)\left(\int dr'' f(r'-r'')\hat{\phi}(r'';\theta,\tau)\right)^3. \tag{6.58}$$

The commutators of $\hat{\phi}$ and $\hat{\pi}$ with $\hat{N}$ yield

$$\frac{\partial}{\partial\theta}\hat{\phi} = r\hat{\pi} + \tau\frac{\partial}{\partial r}\hat{\phi} + \frac{1}{i\hbar}[\hat{\phi},\hat{\Delta}], \tag{6.59a}$$

$$\frac{\partial}{\partial\theta}\hat{\pi} = r\frac{\partial\hat{\pi}}{\partial\tau} - 4G\int dr'(r'-r)f(r'-r)\left(\int dr' f(r'-r'')\hat{\phi}(r'';\theta,\tau)\right)^3$$

$$+ \tau\frac{\partial}{\partial r}\hat{\pi} + \frac{\partial}{\partial r}\hat{\phi} + \frac{1}{i\hbar}[\hat{\pi},\hat{\Delta}], \tag{6.59b}$$

where $\partial\hat{\pi}/\partial\tau$ is given by combining (6.53) and (6.58). We see that the commutator of $\hat{\phi}$ with $\hat{\Delta}$ determines the deviation of $\hat{\phi}$ from being a local field in the presence of the interactions. For the $\hat{\pi}$ equation, (6.59b), the commutator with $\hat{\Delta}$ is augmented by the second term on the right-hand side, which explicitly vanishes in the local interaction limit $f(r'-r) \to \delta(r'-r)$. We are immediately faced with a consistency test since the τ derivative of (6.59a) should yield (6.59b). This requirement is met if we have

$$\left[\hat{\phi}(r;\theta,\tau), \frac{\partial}{\partial\tau}\hat{\Delta}(\theta,\tau)\right] = -i\hbar 4G\int dr'(r'-r)f(r'-r)$$

$$\times \left(\int dr'' f(r'-r'')\hat{\phi}(r'';\theta,\tau)\right)^3. \tag{6.60a}$$

Similarly the τ derivative of (6.59b) must be compatible with (6.59a), and using (6.53) and (6.58) this requirement is met if we have

$$\left[\hat{\pi}(r;\theta,\tau), \frac{\partial}{\partial\tau}\hat{\Delta}(\theta,\tau)\right]$$

$$= -i\hbar\,12G\int dr' f(r'-r)\,dr''(r'-r'')f(r'-r'')\hat{\pi}(r'';\theta,\tau)::$$

$$\times \left(\int dr''' f(r'-r''')\hat{\phi}(r''';\theta,\tau)\right)^2, \tag{6.60b}$$

where the double colon indicates a properly symmetrized product of operators.[35]

[35] The symbol $A:B$ means $1/2(AB+BA)$. The symbol $(A::B^n$ means $1/n$ (AB^n + BAB^{n-1} + B^2AB^{n-2} + ... + B^nA).

Hyperplane-dependent Quantized Fields

The operator $\hat{\Delta}$ is being constrained, and since we have not yet examined the commutation relations between the IHLG generators, we might very well doubt the existence of a suitable $\hat{\Delta}$. However, the commutator between $\hat{H}$ and $\hat{K}$ is trivially satisfied, the commutator between $\hat{N}$ and $\hat{K}$ is satisfied if

$$[\hat{K}, \hat{\Delta}] = 0, \tag{6.61}$$

which is compatible with (6.60), and the commutator between $\hat{N}$ and $\hat{H}$ requires

$$\frac{\partial}{\partial \tau} \hat{\Delta}(\theta, \tau) = -4G \int dr dr'(r-r') f(r-r') \hat{\pi}(r'; \theta, \tau) ::$$
$$\times \left(\int dr'' f(r-r'') \hat{\phi}(r''; \theta, \tau) \right)^3$$
$$= \frac{i}{\hbar} \int dr \left\{ \hat{\pi}(r; \theta, \tau) : [\hat{\pi}(r; \theta, \tau), \hat{\Delta}(\theta, \tau)] + \left(-\frac{\partial^2}{\partial r^2} \hat{\phi}(r; \theta, \tau) \right. \right.$$
$$\left. + \kappa^2 \hat{\phi}(r; \theta, \tau) \right) : [\hat{\phi}(r; \theta, \tau), \hat{\Delta}(\theta, \tau)]$$
$$+ 4G \int dr' f(r-r') [\hat{\phi}(r'; \theta, \tau), \hat{\Delta}(\theta, \tau)] ::$$
$$\left. \times \left(\int dr' f(r-r'') \hat{\phi}(r''; \theta, \tau) \right)^3 \right\}. \tag{6.62}$$

The first equality *implies* (6.60) and is compatible with (6.61)! The second equality is identically satisfied if the commutators on the right-hand side are defined by (6.59) and the equation of motion is used. Thus no *new* information is obtained from the second equality.

For the θ derivative of $\hat{\Delta}$, we have the two equations:

$$\frac{\partial}{\partial \theta} \hat{\Delta}(\theta, \tau) = -\frac{\partial}{\partial \theta} (\hat{N}_{(0)}(\theta, \tau) + \hat{U}_L(\theta, \tau)) \tag{6.63a}$$

and

$$i\hbar \frac{\partial}{\partial \theta} \hat{\Delta}(\theta, \tau) = [\hat{\Delta}(\theta, \tau), \hat{N}], \tag{6.63b}$$

where $\hat{U}_L$ is the explicit part of $\hat{U}$ which will survive the local limit

$$\hat{U}_L(\theta, \tau) = G \int dr \, r \left(\int dr' f(r-r') \hat{\phi}(r'; \theta, \tau) \right)^4. \tag{6.64}$$

The equations, (6.63a and b), will, when the indicated manipulations of the right-hand sides are performed, involve the θ derivatives of $\hat{\phi}$ and $\hat{\pi}$, (6.63a), and the commutators of $\hat{\Delta}$ with $\hat{\phi}$ and $\hat{\pi}$, (6.63b). Using (6.59) and the equation of motion (6.58), we find these two equations, (6.63), to be completely equivalent.

Now it does not appear to be possible to solve the first equality in (6.62) for $\hat{\Delta}(\theta, \tau)$ in terms of $\hat{\phi}$ and $\hat{\pi}$ on the (θ, τ) hyperplane alone,[36] and in a manner compatible with (6.61). On the other hand, it is trivial to solve (6.62) for $\hat{\Delta}(\theta, \tau)$ in terms of the $\hat{\phi}$ and $\hat{\pi}$ on all (θ, τ'') hyperplanes for τ'' lying between some 'initial' value and τ, i.e.

$$\hat{\Delta}(\theta, \tau) = \hat{\Delta}(\theta, \tau') + \int_{\tau'}^{\tau} d\tau'' F[\hat{\pi}, \hat{\phi}; \theta, \tau''], \tag{6.65}$$

where

$$F[\hat{\pi}, \hat{\phi}; \theta, \tau] = -4G \int dr\, dr'(r-r') f(r-r') \hat{\pi}(r'; \theta, \tau) ::$$
$$\times \left(\int dr'' f(r-r'') \hat{\phi}(r''; \theta, \tau) \right)^3. \tag{6.66}$$

If we choose $\tau' = \pm \infty$, then the initial value is unchanged by finite transformations of the IHLG. If we wish to emphasize symmetry of the solution under time-like reversal transformations, we can write

$$\Delta(\theta, \tau) = \frac{1}{2} (\Delta(\theta, \infty) + \Delta(\theta, -\infty)) + \frac{1}{2} \left(\int_{-\infty}^{\tau} d\tau' - \int_{\tau}^{\infty} d\tau' \right) F[\pi, \phi; \theta, \tau']; \tag{6.67}$$

and, finally, if we wish to remove all reference to initial values for $\hat{\Delta}$, so as to obtain a completely explicit expression for $\hat{\Delta}$ in terms of the basic dynamical variables $\hat{\phi}$ and $\hat{\pi}$, we can assume

$$\hat{\Delta}(\theta, \infty) + \hat{\Delta}(\theta, -\infty) = 0 \tag{6.68}$$

for all θ. This yields

$$\hat{\Delta}(\theta, \tau) = \frac{1}{2} \left(\int_{-\infty}^{\tau} d\tau' - \int_{\tau}^{\infty} d\tau' \right) F[\hat{\pi}, \hat{\phi}; \theta, \tau'], \tag{6.69}$$

which relation completes the development of the basic formalism.

At first it may seem as though (6.69) is unacceptable as a basic equation of our model theory, since it requires the complete solution of the τ evolution problem for $\hat{\phi}$ and $\hat{\pi}$ before it can be used. This is not

---

[36] Such a solution, which does *not* satisfy (6.61), is just $\hat{\Delta} = -\hat{N}_{(0)} - \hat{U}_L$.

a flaw of principle here, however, since $\hat{\Delta}(\theta, \tau)$ plays a role only in determining the θ dependence of the basic fields. There is no obstacle, in principle, to solving (6.52) and (6.58) for the τ evolution at a fixed θ for the fields and their commutators on fixed-θ hyperplanes, and then using these solutions to calculate (6.69) and the commutators occurring in (6.59) in order to determine the dependence of the fields on variable θ. In practice, of course, some cyclic iterative procedure must be developed for passing back and forth between the τ evolution and θ evolution problems. This technical problem is presently under study. From the standpoint of the conceptual structure of the model, it is perhaps not surprising that the non-local contribution to the θ derivative of the fields in (6.59)—remember that our expression (6.69) for $\hat{\Delta}$ vanishes in the local interaction limit $f(r-r') \to \delta(r-r')$—depends upon the fields evaluated at all τ values for a given value of θ. An infinitesimal change in θ for fixed τ generates a reoriented hyperplane which, somewhere, intersects *all* hyperplanes, i.e. for *all* τ values, of the original orientation. The consistent interrelations between the various non-local influences that are built into this model serve to emphasize the fundamental role played by the concept of *a point on a hyperplane in space-time*, which here replaces the former conceptual atom of a point in space-time.

REFERENCES

ABERS, E., GRODSKY, I. T., and NORTON, R. E. (1967): *Phys. Rev.* **159**, 1222.
AHARONOV, Y., and ALBERT, D. Z. (1980): *Phys. Rev.* **D21**, 3316.
————(1981): *Phys. Rev.* **D24**, 359.
————(1984a): *Phys. Rev.* **D29**, 223.
————(1984b): *Phys. Rev.* **D29**, 228.
ARDALAN, F., and FLEMING, G. N. (1975): *Journ. Math. Phys.* **16**, 478.
BARUT, A. O., and KLEINERT, H. (1967): *Phys. Rev.* **156**, 1541.
BEBIE, H., and LEUTWYLER, H. (1967): *Phys. Rev. Lett.* **19**, 618.
BJORKEN, J. D., and DRELL, S. (1964): *Relativistic Quantum Fields* (New York: McGraw Hill).
BLOCH, C. (1952): *Dan. Mat. Fys. Medd.* **27** (no. 8).
BLOCH, J. (1967): *Phys. Rev.* **156**, 1377.
BLOCK, Z. (1934): *Phys. Z. Sow. Un.* **5**, 301.
BOYER, C., and FLEMING, G. N. (1974): *Journ. Math. Phys.* **15**, 1007.
CHANG, S. J., and O'RAIFEARTAIGH, L. (1968): *Phys. Rev.* **170**, 1316.
CHRETIEN, M., and PEIERLS, R. E. (1953): *Nuovo Cim.* **10**, 669.
CLAVELLI, L., and HALPRIN, A. (1986): *Lewes String Theory Workshop* (Singapore: World Scientific).

COLLINS, J. (1984): *Renormalization* (Cambridge: CUP).
CORBEN, H. C. (1965): *Phys. Rev. Lett.* **15**, 268.
DIEKS, D. (1985): *Phys. Lett.* **108A**, 379.
DIRAC, P. A. M. (1962): *Rev. Mod. Phys.* **34**, 592.
DYSON, F. J. (1949): *Phys. Rev.* **75**, 486, 1736.
EHLERS, J., HEPP, K., and WEIDENMULLER, H. A. (1973): *Constructive Quantum Field Theory*, Lecture Notes in Physics, vol. 25 (Berlin: Springer-Verlag).
EPSTEIN, H., and GLASER, D. (1973): *Ann. Inst. H. Poincaré* **19**, 211.
FELDMAN, G., and MATTHEWS, P. T. (1967): *Phys. Rev.* **154**, 1241.
FEYNMAN, R. P. (1949): *Phys. Rev.* **76**, 749, 769.
FIERZ, M. (1950): *Helv. Phys. Acta* **23**, 731.
FLEMING, G. N. (1964): *Phys. Rev.* **137**, B188.
—— ——(1966) *Journ. Math. Phys* **7**, 1959.
—— ——(1970): *Phys. Rev.* **D1**, 542.
—— ——(1985): *Towards a Lorentz-Invariant Quantum Theory of Measurement* (Pennsylvania State University preprint), to appear in *Mini-Course and Workshop in Fundamental Physics* (University of Puerto Rico Press).
—— —— and BENNETT, H. (1986): *Hyperplane Dependent Quantum Mechanics in $1+1$ Dimensions* (in preparation).
FRONSDAL, C. (1968): *Phys. Rev.* **171**, 1811.
GALLOVOTTI, G. (1985): *Rev. Mod. Phys.* **57**, 471.
GIOVANNINI, N. (1983): *Helv. Phys. Acta* **56**, 1002.
GLASHOW, S., and GINSPARG, P. (1986): *Physics Today* (May), 7.
GRODSKY, I. T., and STREATER, R. (1968): *Phys. Rev. Lett.* **20**, 695.
HELLWIG, K., and KRAUSE, K. (1970): *Phys. Rev.* **D1**, 566.
ITZYKSON, C., and ZUBER, J. B. (1980): *Quantum Field Theory* (New York: McGraw-Hill).
JAUCH, J. M., and ROHRLICH, F. (1955): *The Theory of Photons and Electrons* (New York: Addison-Wesley).
KATAYAMA, Y., and YUKAWA, H. (1968): *Suppl. Prog. Theor. Phys.* **41**, 1.
KIRZHNITS, D. A. (1966): *Sov. Phys. Uspekhi* **9**, 692.
KRISTENSEN, P., and MØLLER, C. (1952): *Dan. Mat. Fys. Medd.* **27** (no. 7).
LURCAT, F. (1964): *Phys.* **1**, 95.
MAJORANA, E. (1932): *Nuovo Cim.* **9**, 335.
MALIN, S. (1984): *Found. of Phys.* **14**, 1083.
MANDELSTAMM, S. (1962): *Ann. Phys.* **19**, 1.
MAXWELL, N. (1985): *Phil. Sci.* **52**, 23.
MCMANUS, H. (1949): *Proc. Roy. Soc.* **A195**, 323.
NAMBU, Y. (1966): *Suppl. Prog. Theor. Phys.* **37 and 38**, 368.
—— ——(1967): *Phys. Rev.* **160**, 1171.
NEWTON, T. D., and WIGNER, E. P. (1949): *Rev. Mod. Phys.* **21**, 400.
PAIS, A. (1986). *Inward Bound* (Oxford: OUP).
—— —— and UHLENBECK, G. E. (1950): *Phys. Rev.* **79**.

PAULI, W. (1953): *Nuovo Cim.* **10**, 648.
RAYSKI, J. (1954): *Acta. Phys. Polon.* **13**, 95.
REDHEAD, M. L. G. (1983): in *Space, Time and Causality*, ed. R. Swinburne (Dordrecht: Reidel).
ROHRLICH, F. (1974): in *Physical Reality and Mathematical Description*, eds. C. P. Enz and J. Mehra (Dordrecht: Reidel).
SCHWARZ, J. H. (1982): *Phys. Rep.* **89**, 223.
SCHWEBER, S. S. (1961): *Relativistic Quantum Field Theory* (New York: Harper & Row).
―――― (1986): *Rev. Mod. Phys.* **58**, 449.
SCHWINGER, J. (1948a): *Phys. Rev.* **73**, 416.
―――― (1948b): *Phys. Rev.* **74**, 1439.
―――― (1949): *Phys. Rev.* **76**, 790.
SEILER, E. (1982): *Gauge Theories as a Problem of Constructive Quantum Field Theory and Statistical Mechanics*, Lecture Notes in Physics, vol. 5 (Springer-Verlag), 159.
SHIMONY, A. (1986): in *Quantum Concepts in Space and Time* eds. C. Isham and R. Penrose (Oxford: OUP), 182.
STREATER, R., and WIGHTMAN, A. S. (1964): *PCT, Spin and Statistics, and All That*.
STEUCKELBERG, E. C. G., and WANDERS, G. (1954): *Helv. Phys. Acta.* **27**, 667.
TAKABAYASHI, T. (1965): *Suppl. Prog. Theor. Phys.* extra number, 339.
―――― (1966): *Prog. Theor. Phys.* **36**, 185.
TATI, T. (1957): *Prog. Theor. Phys.* **18**, 235.
TELLER, P. (1987); contribution to this volume.
TOMONAGA, S. I. (1946): *Prog. Theor. Phys.* **1**, 27.
VALATIN, J. G. (1951): *Dan. Mat. Fys. Medd.* **26** (no. 13).
WATAGHIN, G. (1934): *Zeit. f. Phys.* **86**, 92.
WIGHTMAN, A. S. (1960): in *Dispersion Relations and Elementary Particles* eds. C. de Witt and R. Omnes (New York: Wiley), 159.
―――― (1962): *Rev. Mod. Phys.* **34**, 845.
WIGNER, E. P. (1939): *Ann. Math.* **40**, 149.
YENNIE, D. R. (1950): *Phys. Rev.* **80**, 1053.
YUKAWA, H. (1950a): *Phys. Rev.* **77**, 219.
―――― (1950b): *Phys. Rev.* **80**, 1047.
―――― (1965): in *Proc. Int. Conf. on Elem. Particles*, (Kyoto), 139.

7
Gauge Theory and the Geometrization of Fundamental Physics

TIAN-YU CAO

Within the twentieth century, two research programmes have sought to describe Nature's fundamental interactions: the geometrical programme initiated by Einstein's general relativity, and the quantum field programme. According to the geometrical programme, interactions are realized through continuous classical fields which are inseparably correlated to geometrical structures of space-time, such as the metric, affine connection, and curvature. This concept of geometrizing interactions has led to a deep understanding of gravitation, but has had no success elsewhere in physics. Indeed, theories of electromagnetic, weak and strong nuclear interactions took a totally different direction, in the late 1920s, with the advent of quantum electrodynamics which initiated the quantum field programme for fundamental interactions. According to this programme, interactions are realized through quantum fields which underlie the local coupling and the propagation of field quanta, but have nothing to do with the geometry of space-time.

The conceptual difference between the two programmes, in ontology and the mechanism for interactions, is so deep that we are bound to ask what is the justification for continuing to talk about the geometrization of fundamental interactions. What I am going to discuss, however, is the possibility of finding an intimate relation between quantum interactions and the geometrical structures of space-time. My central topic will in fact be gauge theory and its geometrical interpretation, but first let me review the relation between gauge theory and quantum field theory, and explain why a proper understanding of quantum field theory cannot be obtained without a proper understanding of gauge theory.

© Tian-Yu Cao 1987.

Recall that quantum fields are localized operator fields. The localized excitation or de-excitation of operator fields comprises the creation or annihilation of discrete field quanta. This novel aspect of quantum fields turns out to have a profound impact on the formulation of the mechanism of interactions in quantum field theory. For these interactions are realized by local coupling among field quanta, meaning the creation or annihilation of field quanta, and the transmission of what are known as virtual quanta. This formulation is thus deeply rooted in the concept of localized excitation of operator fields, via the concept of local coupling. Because of the uncertainty principle, the result of a localized excitation would not be a single momentum quantum, but must be a superposition of all appropriate combinations of momentum quanta. So interactions are described as being transmitted, not by a single virtual quantum, but by a superposition of all appropriate combinations of virtual quanta. But among these virtual quanta, implied by the hypothetical localized excitation, are those seemingly non-physical quanta of arbitrarily high momentum which lead to arbitrarily large contributions, from their interactions with real quanta, to calculations of physical quantities. That is, the well-known divergence difficulty of quantum field theory is unavoidable. Quantum field theory can be considered a consistent theory only when this serious difficulty is overcome, i.e. when it is renormalizable.

It turns out, as theoretical developments within the last two decades have shown, that various theories of quantum fields, especially those describing the fundamental interactions, can be made renormalizable only by introducing the concept of gauge invariance.[1] So for quantum field theory to develop into a successful research programme it is necessary to combine it with gauge theory. Accordingly, the interpretation of quantum field theory is inevitably tied up with the interpretation of gauge theory. To aid this latter process, I turn now to a brief résumé of the circumstances of the original proposal of gauge theory. This had nothing to do with considerations of renormalizability, but developed out of two quite separate trains of thought, one in the 1950s, and an earlier one around 1920.

Modern gauge theory, begun by Yang and Mills in 1954,[2] emerged

---

[1] The relation between gauge invariance and renormalizability is a profound topic. An interesting question is: What is the deeper reason for gauge invariance to be necessary, though not sufficient, for renormalizing a theory?

[2] Yang and Mills (1954a and b).

entirely within the framework of the quantum field programme as follows. More and more new particles were discovered after the Second World War, and various possible couplings among those elementary particles were being proposed. It was therefore necessary to have some principle to choose a unique form out of the many possibilities being considered. The principle suggested by Yang and Mills is based on the concept of gauge invariance, and is hence called the gauge principle.

This idea of gauge invariance orginated in 1918,[3] from the following consideration, due to Weyl: In addition to the requirement of general relativity that coordinate systems have only to be defined locally, so likewise the standard of length, or scale, should only be defined locally. So it is necessary to set up a separate unit of length at every space-time point. Weyl called such a system of unit-standards a gauge system. In Weyl's view, a gauge system is as necessary for describing physical events as a coordinate system. Since physical events are independent of our choice of descriptive framework, Weyl maintained that gauge invariance, just like general covariance, must be satisfied by any physical theory. However, Weyl's original scale invariance was abandoned soon after its proposal, since its physical implications appeared to contradict experiments. For example, as Einstein pointed out, this concept meant that spectral lines with definite frequencies could not exist.

Nevertheless, despite the initial failure, Weyl's idea of a local gauge symmetry survived, and acquired new meaning with the emergence of QM. As is well known, when classical electromagnetism is formulated in the Hamiltonian form, the momentum P_μ is replaced by the canonical momentum $(P_\mu - eA_\mu/c)$. This replacement is all that is needed to fix the form of the electromagnetic interaction. A key concept in QM is the replacement of the momentum P_μ in the classical Hamiltonian by an operator $\left(-i\hbar\dfrac{\partial}{\partial x_\mu}\right)$. So, as Fock observed in 1927,[4] QED could be based on the canonical momentum operator $-i\hbar(\partial/\partial x_\mu - ieA_\mu/\hbar c)$. Soon after that London[5] pointed out the similarity of Fock's work to Weyl's earlier work. Weyl's idea of gauge invariance would then be correct if we replace his scaling factor ϕ_μ with $(-ieA_\mu/\hbar c)$. Thus instead of a scale change $(1 + \phi_\mu dx^\mu)$, we can consider a phase change $(1 - ie/\hbar c A_\mu dx^\mu) \simeq \exp[-ie/\hbar c A^\mu dx^\mu]$, and

---

[3] Weyl (1918a and b).   [4] Fock (1927).   [5] London (1927).

which can be thought of as an imaginary scale change. In 1929 Weyl[6] put all of these considerations together and explicitly discussed the transformation of the electromagnetic potential $(A_\mu \to A'_\mu = A_\mu + \partial_\mu \Lambda)$ and the associated phase transformation of the wave function of a charged particle $\psi \to \psi' = \psi \exp(ie\Lambda/\hbar c)$.

Here the essential clue was provided by the realization that the phase of a wave function could be a new local variable. The previous objections to Weyl's original idea no longer apply because the phase is not directly involved in the measurement of a space-time quantity, as, for example, the length of a vector. Thus in the absence of an electromagnetic field, the amount of phase change can be assigned an arbitrary constant value, since this would not effect any observable quantity. When an electromagnetic field is present, a different choice of phase at each space-time point can then easily be accommodated by interpreting the potential A_μ as a connection which relates phases at different points. A particular choice of phase function will not affect the equation of motion since the phase change and the change in potential cancel each other exactly. In this way the apparent 'arbitrariness' formerly ascribed to the potential is now understood as the freedom to choose any value for the phase of a wave function without affecting the equation. This is exactly what gauge invariance means.

Three points are clear now. First, the correct interpretation of gauge invariance was not possible without quantum mechanics. Second, there is a striking similarity in theoretical structure between gauge theory and general relativity, i.e. the similarity between the electromagnetic field, local phase functions, electromagnetic potential, and gauge invariance on the one hand, and the gravitational field, local tangent vectors, affine connection, and general covariance on the other. Third, as Friedman[7] pointed out, the notion of invariance is different from the notion of covariance: a theory is invariant under a transformation group if some geometrical objects are invariant under the action of the group; a theory is covariant if, under the action of a transformation group, the forms of the equations in the theory are unchanged. Considering this subtle distinction, the gauge invariance here actually should be called gauge covariance, and the symmetry group involved in gauge theory is actually the covariance group, not the invariance group. Incidentally, this remark applies throughout

---

[6] Weyl (1929).   [7] Friedman (1983).

the rest of this paper. Invariance and symmetry are used only in the sense of covariance. For example a theory is said to be symmetrical under the transformations among different representations of relevant observable degrees of freedom if the forms of equations or laws in the theory are unchanged under the action of the transformation group.

The heart of any gauge theory is the gauge symmetry group and the crucial role that it plays in determining the dynamics of the theory. It may be helpful to give some definitions. A symmetry is global if its representation is the same at every space-time point, and local if it is different from point to point. Thus Lorentz symmetry is a global symmetry, and general coordinate transformation symmetry is a local one. Further, symmetry is external if the relevant observable degree of freedom is spatio-temporal in nature, and internal otherwise. Thus Lorentz symmetry is an external symmetry, but the phase symmetry in quantum electrodynamics is an internal one.

Modern gauge theories are just the generalization of local phase symmetry to general localized internal symmetries. The first step was taken by Yang and Mills when they wanted to find the consequences of assuming a law of conservation of isospin which was thought to be the strong interaction analogue of electric charge. The conservation of isospin is a reformulation of the empirical discovery of the charge independence of nuclear force. It assumes, following Heisenberg, that the proton and neutron are but two states in an abstract isospin space of one and the same particle. Now charge conservation is related, in quantum electrodynamics, to phase invariance. So, by analogy, one might assume that the strong interaction would be invariant under isospin rotation. Isospin invariance entails that the orientation of isospin is of no physical significance. The differentiation between a neutron and a proton is then a purely arbitrary process. The arbitrariness, however, is global in character, namely that once a choice is made at one space-time point the choices at all other points are fixed. But as Yang and Mills realized, this is not consistent with the requirement in quantum field theory, that everything be defined locally.

To get a local isospin invariant theory, Yang and Mills were guided by an analogy to quantum electrodynamics, and introduced a gauge potential B_μ, which under an isospin rotation transforms as $B_\mu \rightarrow B'_\mu = S^{-1} B_\mu S + \frac{i}{\varepsilon} S^{-1} \frac{\partial S}{\partial x_\mu}$, in order to counteract the space-time dependence of the isospin rotation S of a nucleon wave function. Here $S = \exp(-i\theta(x)L)$ is an element of the local isospin group, a non-Abelian Lie group: L has three components L_i which are the generators of the group. The results they obtained are significant.

First, it is easy to show that the potential B_μ must contain a linear combination of $L_i: B_\mu = 2b_\mu \cdot L$. This relation explicitly displays the dual role of the Yang–Mills potential as both a four-vector field in space-time and an Su(2) isospin vector operator, namely that Yang–Mills potentials are non-Abelian fields. Generalization to other non-Abelian fields when other internal quantum numbers from isospin, such as Su(3) colour, are considered is conceptually straightforward.

Second, the field equations satisfied by the twelve independent components b_μ^i of the potential B_μ, and their interaction with any field having an isospin charge, are essentially fixed by the requirement of isospin invariance in the same way that the electromagnetic potentials A_μ and their interaction with charged fields are essentially determined by the requirement of phase invariance. Thus we have a powerful principle for determining the forms of the fundamental interactions, namely the gauge principle. This principle, absent from the older quantum field theory, is the corner-stone of the gauge field programme.

According to the gauge field programme, all kinds of fundamental interactions can be described by the gauge potentials. So the problem of geometrizing fundamental physics can be reduced to the problem of geometrizing gauge theory.

Before discussing this central point, let me pause to examine the universality of the gauge principle. The claim that the gauge principle is universally applicable is challenged when external symmetry is involved. There are two questions: (i) Is the gravitational interaction dictated by the external symmetry, that is by the principle of general covariance? (ii) Can we gauge an external symmetry to get a quantum of gravity?

As early as 1917, Einstein's original claim that general covariance, together with the equivalence principle, leads to general relativity was criticized by Kretschmann.[8] It has also been challenged later on by Cartan, Wheeler[9] and Friedmann,[10] among others. These critics

[8] Kretschmann (1917).   [9] Misner, Thorne, and Wheeler (1973).
[10] Friedman (1983).

argue that the principle of general covariance is of no physical content whatever since all non-general-covariant theories can always be recast in a general covariant form. The only thing we should do is just to replace ordinary derivatives by covariant derivatives, adding an affine connection into the theory and saying that there exist coordinate systems in which the components of the affine connection Γ^k_{ij} happen to be equal to zero and the original formalism is valid.

But it seems to me that the gravitational interaction is actually dictated by a local external symmetry. For these critics admit that, in the reformulations of non-general covariant theories, $\Gamma^k_{ij} = 0$, but this implies that space-time is flat and that the external symmetry is global. However, true general covariance is equivalent to a local symmetry, the only type of symmetry that can be satisfied by the curved space-time in which gravity can be accommodated. Hence the supposed general covariance of these reformulated theories is spurious.

The second question is also not without controversy. Many attempts have been made, different symmetry groups being tried, to find a quantum theory of gravity. Among these, the work by Hehl and his collaborators[11] seems to prove convincingly that the Poincaré group, i.e. the group of Lorentz transformations and translations, can lead to the most plausible gauge theory of gravity provided it is interpreted actively. Now all earlier attempts based on the local Poincaré group failed to be renormalizable. About ten years ago, however, a more powerful symmetry took the stage, i.e. supersymmetry. Local supersymmetry leads to the theories of supergravity, which were believed to be able to unify gravity with the other interactions.[12] Despite early optimism that incorporating supersymmetry would render supergravity renormalizable, or perhaps even finite, it now seems probable that none of the supergravity theories will lead to a consistent quantum theory. Then, about five years ago, came superstring theory[13]; it takes supergravity as a low-energy approximation, and escapes from the divergence difficulty if space-time has ten dimensions. It is too early to say whether superstring will provide a succcssful theory; in any case, it seems not impossible to get a gauge theory of gravity.

---

[11] Hehl *et al.* (1976).
[12] Van Nieuwenhuizen (1981).
[13] Green and Schwarz (1981, 1985).

Now let me turn to the central topic, the geometrization of fundamental physics. The starting-point here is the geometrization of gravity: making Poincaré symmetry local removes the flatness of space-time and requires the introduction of some geometrical structures of space-time, such as metric, affine connection, and curvature, which are correlated with gravity.

For other fundamental interactions, which, it is believed, can be described as gauge interactions, we find that the theoretical structures of the corresponding theories are exactly parallel to that of gravity. There are internal symmetry spaces: phase space, for electromagnetism, which looks like a circle; isospace, which looks like the interior of a three-dimensional sphere; colour space for strong interaction, etc. The internal space defined at each space-time point is called a fibre, and the union of this internal space with space-time is called fibre-bundle space. Then we find that the local gauge symmetries remove the 'flatness' of the fibre-bundle space since we assume that the internal space directions of a physical system at different space-time points are different. So the local gauge symmetry also requires the introduction of gauge potentials, which are responsible for the gauge interactions, to connect internal directions at different space-time points. We also find that the role the gauge potentials play in fibre-bundle space in gauge theory is exactly same as the role the affine connection plays in curved space-time in general relativity. Then why can't we speak of the geometrization of gauge interactions, if we accept that gravitational interaction is geometrized because it is correlated with the geometrical structures of curved space-time, namely the affine connection?

A critic might reply, however, that this is not a genuine geometrization, since the geometrical structure in gauge theories is only defined on the fibre-bundle space, and the latter is only a kind of mathematical structure, totally different from real space-time. The bundle manifold here is just a trivial product of space-time and internal space, two factor spaces having nothing to do with each other. So the so-called geometrization would just be meaningless rhetoric.

At first glance the point at issue here seems to lie only in different interpretations of what should be called genuine geometrization. But I claim this is not the case, and that further consideration of this point, far from a mere terminological dispute, has far-reaching potential to deepen our understanding of fundamental physics.

To support this claim, I shall start further discussion with a simple question: what is geometry? Before special relativity, the only physical geometry was Euclidean geometry in three-dimensional space; other geometries, though logically possible, were viewed as only a kind of mental fiction. With special relativity came Minkowskian geometry in four-dimensional space-time. No one today would deny that this is a genuine geometry. An important reason for this belief is that the four-dimensional space-time on which the Minkowskian geometry is defined is not just a trivial product of three-dimensional space and one-dimensional time; rather, the Lorentz rotation mixes the spatial and temporal indices. Riemannian geometry in four-dimensional space-time is different again from Minkowskian geometry. Is this also a genuine geometry, or just a kind of mathematical trick for describing gravitation against the background of Minkowskian geometry?

One of the lessons general relativity teaches us is that the existence of gravitational interaction is of necessity reflected in the geometrical character of space-time. Even if we start with Minkowskian geometry and reformulate general relativity in the form of a quantum field theory in which gravity is described by a massless spin-2 field, the resultant theory, as Deser[14] showed, always involves a curved metric, not the originally postulated flat metric. The relation between flat and curved metrics here is quite similar to the relation between bare and dressed charges in quantum field theory. So Riemannian geometry should be viewed as a genuine geometry of space-time. Or, equivalently, we should recognize that the character of the genuine geometry of space-time is actually affected by gravity. This is precisely what Riemann speculated in his famous inaugural lecture and what Einstein insisted all throughout his life.

Then what is the situation when internal symmetries are concerned, or equivalently, when interactions other than gravitation are concerned? Before going into this problem, let me first outline the further evolution of ideas of geometry after general relativity.

In 1921, Eddington introduced an idea about 'the geometry of the world-structure, which is the common basis of space and time and things'.[15] In this broadest sense, Eddingtonian geometry would contain elements correlated to matter, and to interaction mechanisms of material systems. Such a kind of geometry has been explored by

[14] Deser (1970). [15] Eddington (1921).

Einstein and Schrödinger in their unified field theories, and by Kaluza and Klein in their five-dimensional relativity.[16] And as I will explain, interest in Eddingtonian geometry has been reviving in the last decade in the fibre bundle version of gauge theory,[17] in supergravity,[18] in modern Kaluza–Klein theory[19], and in superstring theory.[20]

In the original Yang–Mills theory, the so-called differential formalism, internal degree of freedom, though generalized from the simple phase of complex numbers to the generators of a Lie group, had nothing to do with space-time. They just generalized quantum electrodynamics, knowing nothing of its geometrical meaning. But in 1967 a contact point between internal and external degrees of freedom emerged. T. T. Wu and C. N. Yang in their 1967 paper[21] suggested a solution to isospin gauge field equations: $b_{i\alpha} = \Sigma_{i\alpha\tau} f(r)/r$, where $\alpha = 1,2,3$, designates isospin-index, i and τ are space-time indices, and r is the length of the three-vector (x_1, x_2, x_3). This solution explicitly mixes isospin and space indices. Conceptually, this was a significant development. If Minkowskian geometry can be regarded as a genuine geometry simply because spatial and temporal indices are mixed by Lorentz rotation, then the Wu–Yang solution had obviously provided a wider basis for a new genuine geometry.

A further development in this direction was the monopole solution of an Su(2) gauge theory suggested independently by 'tHooft and by Polyakov in 1974.[22] In this case, the monopole solution relates directions in physical space with those in internal space. The same type of correlation happens also in the instanton solutions of gauge theory.

The geometrical implication of gauge theory become discernible only with the suggestion, by Yang in 1974,[23] of an integral formalism for gauge fields. The essential point of the integral formalism is the introduction of the concept of a non-integrable phase factor. A non-integrable phase factor is defined as $\phi_{A(A+dx)} = I + b_\mu^k(x) X_k dx^\mu$, where $b_\mu^k(x)$ are the components of a gauge potential, and X_k are the

[16] Kaluza (1921); Klein (1926, 1927).
[17] Daniel and Viallet (1980).
[18] Van Nieuwenhuizen (1981).
[19] Hosotani (1983).
[20] Green and Schwarz (1981, 1985).
[21] Wu and Yang (1975).
[22] 'tHooft (1974); Polyakov (1974).
[23] Yang (1974).

generators of the gauge group. With the help of this concept, Yang found that (i) the parallel displacement concept of Levi-Civita was a special case of a non-integrable phase factor with the gauge group GL (4); (ii) linear connection was a special case of gauge potential; and (iii) Riemannian curvature was a special case of gauge field strength. Although Yang's specific GL (4) gauge theory of gravity was criticized by Pavelle[24] for differing from Einstein's theory and for possible conflict with observation, these general ideas of Yang are now universally accepted. This work was impressive in showing that the concept of gauge fields is deeply geometrical. This impression was strengthened by the work on global formulations of gauge fields, by Wu and Yang in 1975.[25] Here the global geometrical connotations of gauge fields including those implied by the Aharanov–Bohm effect and the 'tHooft–Polyakov monopole, were explored and formulated in terms of fibre-bundle concepts.

A fibre bundle over R^4 is a generalization of a product space of space-time and an internal space, which allows for possible twisting in the bundle space and, therefore, gives rise to a non-trivial fusion of space-time with internal space. Mathematically the external and internal indices can be blended together by certain transition functions whose role in gauge theory is played by the generalized phase transformations.[26]

Given the above developments culminating in the fibre-bundle version of gauge theory, it is clear now that parallel to the external geometry associated with spatio-temporal degrees of freedom, there is an internal geometry associated with internal degrees of freedom. The two types of geometry can be put in the same mathematical framework. However, despite the impressive mathematical similarity and more impressive possibility of blending external with internal degrees of freedom, the real picture of the unification of the two types of geometry, which is correlated with the unification of gravitational and other fundamental interactions, is still quite vague.

Now let me turn to more recent developments in fundamental physics: supersymmetry and supergravity,[27] modern Kaluza–Klein theory[28], and superstring theory.[29] Supersymmetry is a symmetry

[24] Pavelle (1975).
[25] Wu and Yang (1975).
[26] Daniel and Viallet (1980).
[27] Van Nieuwenhuizen (1981).
[28] Hosotani (1983).
[29] Green and Schwarz (1981, 1985).

that relates bosons to fermions. In quantum field theory boson fields have dimension one and fermion fields have dimension 3/2 in order that the action be dimensionless (in units $\hbar = c = 1$). The reason is that boson fields have two derivatives in their action while fermion fields have only one. It is not difficult to see that two supersymmetry transformations will certainly lead to a gap of one unit of dimension. The only dimensional object different from fields themselves available to fill this gap is the derivative. Thus in any global supersymmetry model we can always find a derivative appearing in a double transformation relation, purely on dimensional grounds.

Mathematically therefore the global supersymmetry resembles taking the square root of the translation operator. So actually it is not an internal symmetry but an enlargement of the Poincaré group to what is called the 'super-Poincaré group'. This amounts to an extension of space-time to superspace that includes extra spinorial anticommuting coordinates as well as ordinary coordinates.

In local theory the translation operator differs from point to point. This is precisely the notion of a general coordinate transformation and leads us to expect that gravity must be present. Indeed, guided by the requirement of local supersymmetry invariance and using 'Noether's method', we can actually get massless spin-3/2 field gauging supersymmetry, i.e. the gravitino, and massless spin-2 field gauging space-time symmetry, i.e. the graviton. So the local gauge theory of supersymmetry implies a local gauge theory of gravity. This is the reason for such a local supersymmetry theory being called supergravity.

In simple supergravity, the number of supersymmetry generators N is equal to one, if N is larger than one, then we get the extended supergravity. One of the special features of the extended supergravity theories is that they have, in addition to the space-time symmetries related to gravity, a global U(N) internal symmetry which relates all the particles of the same spin and hence has nothing to do with supersymmetry. It was proved that the internal symmetry could be made local so that the non-gravitational interactions could be incorporated. Great excitement followed the discovery in the $N = 8$ model of an extra local Su(8) symmetry in the hope that the local Su(8) group could produce the spin-1 and spin-1/2 bound states which are needed for grand unified theories. In spite of all these developments, however, the relation between supersymmetry and extra internal symmetry is still unclear at present. So the relation between gravitational and other interactions, and the relation between external and internal geometries are still vague within the context of supergravity.

The next theory I want to discuss is the modern Kaluza–Klein theory. In the original Kaluza–Klein theory, an extra dimension of space, compactified in low-energy experiments, was grafted onto the known four-dimensional space-time, in order to accommodate electromagnetism. In the revitalized Kaluza–Klein theory, the number of extra space dimensions become seven, taking into account the number of symmetry operations embodied in grand unified theories and extended $N = 8$ supergravity. It has been supposed that the seven extra space dimensions are compactified in low energy as a hypersphere. The seven-sphere contains many additional symmetries which are intended to model the underlying gauge symmetries of the force fields. This means that the internal symmetries are the manifestation of the geometrical symmetries associated with the extra compactified space dimensions, and that all the kinds of geometries associated with internal symmetries are genuine space geometries, i.e. the geometries associated with extra space dimensions.

In this case the question of what is the relation between internal and external geometries seems to have turned out to be quite irrelevant. But in a deeper sense the question remains profound. What is the real issue when we talk about the geometrization of gauge field theory? The geometrization thesis only makes sense if the geometrical structures in four-dimensional space-time are actually correlated to the gauge interactions other than gravity or supergravity, or equivalently, if they are mixed with the geometrical structures associated with extra dimensions. We shall not know whether this is the case or not until the hypothetical compactification scale, the Planck scale $T = 10^{38}$ GeV or so, is reached. So it is still an open question at present. Nevertheless, the modern Kaluza–Klein theory does open a door for establishing the correlation between non-gravitational gauge potentials and the geometrical structures in four-dimensional space-time via the geometrical structures in extra dimensions. Within this theoretical context, the geometrization claim, related to, though not identified with, the unification of gravity and other gauge interactions, is in principle testable, and cannot be accused of being nonfalsifiable or irrelevant to the future development of fundamental physics.

The last development in fundamental physics which is relevant to this paper is superstring field theories, which have evolved since 1980 from the spinning string model of hadrons originally developed in the early 1970s. In the string model a single quantum string, different from a structureless point particle, has an infinite number of states with masses and spins which increase without limit; the scale for the mass splitting between these states is set by the string tension T. In superstring theories which incorporate supersymmetry, the ground states have zero mass and are in fact simply the fundamental states of supergravity point field theories. Superstring theories are therefore interpreted as theories that include gravity and the natural mass scale set by T is the Planck scale, i.e. $T = 10^{38}$ GeV.

Just as in the original spinning string model, superstrings also require ten-dimensional space-time. So, to be of relevance to physics, the extra six dimensions must compactify and be very small. This is analogous to the idea employed in the Kaluza–Klein version of supergravity point field theory. However, while in the context of supergravity point field theory the extra dimensions make the divergence problem worse than they already are in four dimensions, some quantum superstring field theories seem to have no infinities at all. The crucial point here is that there is a cancellation mechanism provided by the higher mass states, absent in point field theories, which are excited at energies approaching the Planck scale, at which the extra dimensions compactify.

In ten-dimensional quantum superstring theories there are gravitational and Yang–Mills anomalies, that is the violation of the conservation of the Yang–Mills charges and the energy-momentum. Requiring the absence of all anomalies leads to requiring a very intimate relationship between gravitational and Yang–Mills interactions. For example in the low-energy approximation of one type of superstring theory, the existence of a Yang–Mills sector requires the existence of the gravitational sector, and this unification of the Yang–Mills and gravitational interactions leads to the relationship $K = \text{const } g^2 T$ between the gauge coupling constant g and the gravitational coupling K. In another type of theory, the relation is $K = \text{const } g/T^{1/2}$. Such a direct relation between Yang–Mills interactions and gravity is absent in supergravity point field theories, and the interplay between gravity and Yang–Mills interactions is the key to understanding how the anomalies cancel.

Another aspect of the unification of Yang–Mills and gravity can be found in the geometrical constraint of the theory: $\int (\text{Tr} R_{\mu w} R_{\rho \sigma} - 1/30 \text{Tr} F_{\mu \omega} F_{\rho \sigma}) \Sigma^{\mu \omega \rho \sigma} = 0$, where R is the Riemann curvature tensor,

F is the Yang–Mills strength, and the integral is over an arbitrary four-dimensional submanifold of the full ten-dimensional space-time. This is a direct relation between geometrical properties of ten-dimensional space-time and of the Yang–Mills field configurations.

One of the most remarkable features of the superstring theories which distinguishes them from the usual Kaluza–Klein schemes is that the Yang–Mills group in the ten-dimensional theory provides all the experimentally desirable internal symmetries, the dimensional compactification produces no extra gauge invariance, while in the usual Kaluza–Klein schemes the gauge symmetries are supposed to emerge from that of the compactified internal space. For example in the eleven-dimensional supergravity theory all the observed four-dimensional Yang–Mills symmetries are supposed to emerge from the symmetries of the compactified seven-dimensional space.

From the above brief review, we find there are three versions of geometrization of non-gravitational gauge interactions:

1. Fibre-bundle version, in which the gauge interactions are correlated with the geometrical structures of internal space. Since it is possible to get a non-trivial fusion of space-time with internal space, the gauge interactions also have some indirect relation with space-time geometry. But the essence of the internal space is still a vexing problem: Is it a physical reality as real as space-time, or just a mathematical structure?

2. Kaluza–Klein version, in which extra space dimensions which compactify in low-energy experiments are introduced and the gauge symmetries by which the forms of gauge interactions are fixed are just the manifestation of the geometrical symmetries of the compactified space. Here the mediator between the gauge interactions and the space-time geometry is no longer the vexing internal space but the real though compactified extra space dimensions. The assumption of the reality of the compactified space is substantial and is in principle testable, although its *ad-hoc*-ness makes it difficult to differentiate it from the internal space in the fibre-bundle version.

3. Superstring version, in which the introduction of extra compactified space dimensions is due to different considerations from just reproducing the gauge symmetry. Therefore, the properties and structures of the compactified dimensions are totally different from those in the Kaluza–Klein version. For example there is no symmetry in the compact dimensions from which the gauge symmetries emerge; the gauge interactions are correlated with the geometrical structure of

ten-dimensional space-time as a whole and not just with the extra dimensions. Besides, in the superstring version, there could be some direct and testable relations between gravitational and non-gravitational gauge interactions.

So in my opinion, the thesis of geometrization of fundamental interactions is of both scientific and philosophical significance. It raises the question of the relation of space-time to all kinds of fundamental interactions, not just to gravity. If our knowledge of dynamic properties and structures of curved space-time is possible only with Einstein's general relativity, which relates space-time with gravity, then the geometrization thesis will certainly provide the new possibility to widen our knowledge and deepen our understanding of space-time. It predicts the existence of higher-dimensional space-time and some of its properties and structures. It establishes some direct relation between gravitational and non-gravitational interactions, etc.

Then what is its significance for our understanding the essence of quantum gauge interactions? The reply would be: Exactly the significance of general relativity for our understanding the essence of gravitation!

REFERENCES

DANIEL, M., and VIALLET, C. M. (1980): 'The Geometrical Setting of Gauge Theories of the Yang–Mills Type', *Rev. Med. Phys.* **52**, 175.

DESER, S. (1970): 'Self-Interaction and Gauge Invariance', *Gen. Rel. Grav.* **1**, 9.

EDDINGTON, A. S. (1921): 'A generalization of Weyl's Theory of the Electromagnetic and Gravitational Fields', *Proc. Roy. Soc.* **A99**, 104.

FOCK, V. (1927): 'Über die invariante from der Wellen- und der Bewegungsgleichungen für einen geladenen Massenpunkt', *Z. Phys.* **39**, 226.

FRIEDMAN, M. (1983): *Foundations of Space-time Theories* (Princeton, NJ: Princeton University Press).

GREEN, M. B., and SCHWARZ, J. H. (1981): 'Supersymmetrical Dual String Theory', *Nucl. Phys.* **B181**, 502.

—— —— (1985): 'Infinity Cancellations in SO (32) Superstring Theory', **B151**, 21.

HEHL, F. W. et al. (1976): 'General Relativity with Spin and Torsion: Foundations and Prospects', *Rev. Mod. Phys.* **48**, 393.

HOSOTONI, Y. (1983): 'Dynamical Gauge Symmetry Breaking as the Casimir Effect', *Phys. Lett.* **129B**, 193.

KALUZA, T. (1921): 'Zum Unitätsproblem der Physik', *Sitzber. Preuss. Akad. Wiss.* 966.

KLEIN, O. (1926): 'Quantemtheorie und fündimensionale Relativitätstheorie', *Z. Phys.* **37**, 895.
—— (1927): 'Zur fündimensionalen Darstellung der Relativitästheorie', *Z. Phys.* **46**, 188.
KRETSCHMANN, E. (1917): 'Über den Physikalischen Sinn der Relativitätspostulaten', *Ann. Phys.* **53**, 575.
LONDON, F. (1927): 'Quantenmechanische Deutung der Theorie von Weyl', *Z. Phys.* **42**, 375.
MISNER, C. W., THORNE, K. S. and WHEELER, J. A. (1975): *Gravitation* (San Francisco, Calif: Freeman).
PAVELLE, R. (1975): 'Unphysical Solutions of Yang's Gravitational-Field Equations', *Phys. Rev. Lett.* **34**, 1114.
POLYAKOV, A. M. (1974): 'Particle Spectrum in Quantum Field Theory', *JETP Lett.* **20**, 194.
'T HOOFT, G. (1974): 'Magnetic Monopoles in Unified Gauge Theories', *Nucl. Phys.* **B79**, 276.
VAN NIEUWENHUIZEN, P. (1981): 'Supergravity', *Phys. Lett. Rep.* **68**, 189.
WEYL, H. (1918a): *Space-Time-Matter*. Trans. by H. L. Brose (London: Methuen, 1922).
—— (1918b): 'Gravitation und Elektrizität', *Sitzber. Preuss. Akad. Wiss*, 465.
—— and MILLS, R. L. (1954a): 'Isotopic Spin Conservation and a Generalized Gauge Invariance', *Phys. Rev.* **95**, 631.
—— —— (1954b): 'Conservation of Isotopic Spin and Isotopic Gauge Invariance', *Phys. Rev.* **96**, 191.
—— (1929): 'Elektron und Gravitation', *Z. Phys.* **56**, 330.
WU, T. T. and YANG, C. N. (1967): 'Some Solutions of the Classical Isotopic Gauge Field Equations', in *Properties of Matter under Unusual Conditons*, H. Mark and S. Fernbach, (eds.) (New York: Wiley-Interscience), 349–354.
—— —— (1975): 'Concept of Nonintegrable Phase Factors and Global Formulation of Gauge Fields', *Phys. Rev.* **D12**, 3845.
YANG, C. N. (1974): 'Integral Formalism for Gauge Fields', *Phys. Rev. Lett.* **33**, 445.

IV
Mathematical Foundations of Quantum Field Theory

8
Why Should Anyone Want to Axiomatize Quantum Field Theory?
RAY F. STREATER

Quantum field theory was really invented by Dirac in 1927. He considered classical electromagnetic fields in a cubical box with periodic boundary conditions. This theory is described by a collection of independent classical harmonic oscillators, which Dirac quantized as in the non-relativistic theory. Physical quantities are then ultimately obtained by taking the limit, as the size of the box goes to infinity, of a corresponding quantity in the boxed theory. Dirac's procedure leads to the prediction that light is composed of quanta which obey Bose–Einstein statistics.[1] These days, the relativistic free fields are rather easily[2] quantized in a rigorous way without first quantizing in a box; but in constructing[3] fields with interaction we must still use a box and other, much more elaborate approximations.

Two of the motivating arguments that led Dirac to his relativistic wave equation for matter are now regarded as misguided. When Schrödinger wrote down his non-relativistic (better called Galilean-invariant) wave equation, he also wrote down the relativistic, i.e. Lorentz-invaiant, version

$$\frac{1}{c^2}\frac{\partial^2 \phi}{\partial t^2} - \frac{\partial^2 \phi}{\partial n^2} - \frac{\partial^2 \phi}{\partial y^2} - \frac{\partial^2 \phi}{\partial z^2} + m^2 \phi = 0$$

abbreviated to

$$(\Box + m^2)\phi = 0.$$

This is now known as the complex Klein–Gordon equation. But Schrödinger and others rejected it on several grounds. One was that it

---

[1] P. A. M. Dirac, *Principles of Quantum Mechanics* (Oxford: OUP, 4th edn. 1958).
[2] I. E. Segal, *Mathematical Problems of Relativistic Physics* (American Mathematical Society, 1971).
[3] J. Glimm and A. Jaffe, *Quantum Physics* (Berlin: Springer-Verlag, 1981).

© R. F. Streater 1987. Based on a talk given to the Department of History and Philosophy of of Science, King's College, London, in 1985.

admits solutions of negative energy, as well as the wanted solutions of positive energy. This is in contrast to the Schrödinger equation, all of whose solutions are of positive energy. Here we see the principle of *positive energy* playing a role in formulating the theory; this is the beginning of the axiomatic method.

Nowadays it seems obvious to a mathematician that we can obtain a theory with positive energy by considering the set of complex ϕ obeying

$$\left.\begin{array}{r}(\Box + m^2)\phi(x, y, z, t) = 0 \\ P_-\phi = 0\end{array}\right\},$$

where P_- is the projection operator onto states of negative energy. The second equation states that ϕ has no component of negative energy. If Schrödinger thought of this, he rejected it; P_- is a non-local operator with the remarkable property that if ϕ vanishes in some open region of space-time but is not the zero wave function, then $P_-\phi \neq 0$. Thus no positive-energy solutions exist that vanish in space-time regions, and so one cannot find localized solutions obeying $P_-\phi = 0$. The proof of this uses no more than the theory of Laplace transforms: Let P_- be the projection operator onto the subspace of solutions to $(\Box + m^2)\phi(x, y, z, t) = 0$, such that the Fourier transform

$$\frac{1}{\sqrt{(2\pi)}}\int e^{-iEt}\phi(x, y, z, t)dt = \hat{\phi}(x, y, z; E)$$

is zero for $E > 0$. Let $P_+ = 1 - P_-$; then the condition $P_-\phi = 0$ becomes $P_+\phi = \phi$ and this is easily shown to imply that

$$\phi(x, y, z, t) = \frac{1}{\sqrt{(2\pi)}}\int_0^\infty e^{iEt}\hat{\phi}(x, y, z, E)\,dE$$

by Fourier's inversion theorem. The integral on the right-hand side converges also if t is complex, provided that t has positive imaginary part. Thus, a positive-energy solution is the boundary value, for real t, of an analytic function of t, and so cannot vanish in any interval without being zero (by the principle of uniqueness of analytic continuation and Schwarz's reflection principle). It is in this sense that P_- is non-local.

These non-local properties of $P_-\phi = 0$ might have been intuitively understood by Schrödinger or Dirac, who would have felt that all decent equations should be differential equations; if so, we see *locality* playing a role in model-building.

Another trouble thought to belong to the relativistic wave equation was that the four-vector

$$j^\mu = \frac{1}{2i}(\phi^*\partial_\mu\phi - \phi\partial_\mu\phi^*), \quad \mu = 0, 1, 2, 3$$

does not give a *positive-definite* probability current. That is the conserved quantity

$$Q = \int d^3x \frac{1}{2i}(\phi^*\dot\phi - \phi\dot\phi^*)$$

is not positive, unlike the similar quantity in Schrödinger's theory. We now interpret this as the electric charge of the meson, and this does not need to be positive. There *is* a positive-definite quantity, or Hilbert space norm, which is conserved in time.[4] This is written in terms of the wave function at a given time as follows. Let $\phi(x, y, z, t)$ be a positive-energy solution to the Klein–Gordon equation, and let $f(x, y, z)$, $g(x, y, z)$ be the *Cauchy data* of its real part, at $t = 0$:

$$f(x, y, z) = \operatorname{Re}\phi(x, y, z, 0)$$
$$g(x, y, z) = \operatorname{Re}\dot\phi(x, y, z, 0).$$

Then define $\|\phi\|^2 = \int d^3x\left(g\frac{1}{\omega}g + f\omega f\right)$, where ω is the pseudo-differential operator

$$\omega = \sqrt{(-\Delta + m^2)}.$$

Schrödinger almost certainly knew of this scalar product in its more usual form:

$$\int |\varphi(p_1, p_2, p_3)|^2 \frac{d^3p}{\omega_p},$$

where $\theta(p^0)\delta(p^{0\,2} - \mathbf{p}^2 - m^2)\varphi(\mathbf{p})$ is the four-dimensional Fourier transform of ϕ. In either form, the norm is non-local, in that two-wave functions, non-zero in two disjoint sets at $t = 0$, are not orthogonal, however far apart the two sets are. If we interpret $|\phi(\mathbf{x}, t)|^2$ as the probability density, at time t, that the particle is at $\mathbf{x}$, then it has a finite (though small) probability of being found far away. Schrödinger would have rejected this 'non-locality'.

All these troubles are, for the free field, solved by second

---

[4] P. J. M. Bongaarts, 'Linear Fields according to I. E. Segal' in *Mathematics of Contemporary Physics*, ed. R. F. Streater (New York: Academic Press, 1971).

quantization, of which more later. If Schrödinger had done this, then Dirac might not have discovered his equation.

The Dirac equation

$$\left(i\frac{\partial}{\partial x^\mu}\gamma^\mu + m\right)\psi(\mathbf{x}, t) = 0,$$

where $\psi = (\psi_1, \psi_2, \psi_3, \psi_4)$ is a four-component complex spinor, and $\gamma^\mu, \mu = 0, 1, 2, 3$, are four 4×4 matrices defining the Dirac algebra, is of the first order in the differential $\frac{\partial}{\partial t}$. A solution ψ is therefore determined by the value of ψ at an initial value, say $t = 0$. We say[5] that a system obeys the law of *primitive causality* if the state of the system at $t = 0$ determines the state at a later time. It is amazing that even today some textbooks argue[6] that this property makes the Dirac equation preferable to the Klein–Gordon equation, which is of second order. For, any second-order equation can easily be written as a system of first-order equations. Thus, the Klein–Gordon equation also obeys primitive causality when we specify the state at $t = 0$ by the *pair* of functions $(\phi, \dot\phi)$, the Cauchy data of the solution. Thus, although primitive causality is a good property to require, and might have influenced Dirac in his rejection of the Klein–Gordon equation, its use was a mistake.

Dirac's theory has a positive, Lorentz-invariant local norm

$$\sum_{\alpha=1}^4 \int |\psi_\alpha(\mathbf{x}, 0)|^2 d^3x,$$

which is the space integral of the time-component of a conserved four-vector probability current. This 'solved' the problems of non-locality or non-positivity of the Klein–Gordon theory, and was highly regarded by experts at the time. However, it still has negative-energy solutions, which Dirac hoped would remain uncoupled. But the projection operator P_- onto negative-energy states is still non-local. Moreover, when coupled to the electromagnetic field, these negative-energy states could show up (the Klein paradox).

Dirac hoped to exclude the negative-energy states by a remarkable assumption: in the vacuum state, all these energy levels are occupied; thus the vacuum is envisaged as a 'sea' of negative-energy electrons.

---

[5] R. Haag and D. Kastler, 'An Algebraic Approach to Quantum Field Theory', *J. Math. Phys.* **5** (1964), p. 848.

[6] Dirac, *Principles of Quantum Mechanics*.

Dirac hoped that the Pauli exclusion principle, which prohibits two particles from occupying the same state, would ensure that the negative-energy states would never be observed in his theory, since no physical particle could make a transition to a filled state. He noticed, however, that any of the negative-energy electrons could be excited and knocked out to a positive-energy state, leaving a hole. This hole behaves as a positively charged particle of the same mass. At first, scientists, Weyl included, tried to identify this as the proton, but when positrons were discovered, a great boost was given to Dirac's hole theory. But why cannot we measure the electric field of the infinite number of electrons in the sea? This and all remaining problems were solved when Dirac's hole theory was reformulated through second quantization: ψ_α becomes an operator which like the quantized electromagnetic field of Dirac, creates and annihilates particles of positive energy. The negative-energy part ψ_- of ψ *removes* energy. The positive-energy condition is then $\psi_-\Omega = 0$, where Ω is the vacuum state. By quantizing ψ with anticommutators, electrons automatically obey Fermi–Dirac statistics and therefore the Pauli exclusion principle. The final problem, the Klein paradox, was finally understood in the 1970s with the advent of anomalies.

The second quantization of the Klein–Gordon equation, where ϕ becomes an observable rather than a state, also bypasses the non-locality which plagues it in the 'one-particle' interpretation. By 1936 the relativistic free quantized fields describing photons, electrons, protons, and charged and neutral particles of spin 0 were available. These are *relativistic*, act in Hilbert spaces of states with *positive transition probabilities*, are *local* (defined as averages over arbitrarily small regions), have *no states of negative energy*, and obey *primitive causality* and *local commutativity*. The latter means that observables commute when localized in spacelike separated regions. Primitive causality is obeyed in a strong sense: the field in a space-time region is a function of the fields in its domain of dependence (points connected by some backward time-like four-vector to the region). This theory is a theory of observables, in which the field is primary and the properties of the states and the particles are derived, not postulated. So field-theorists were born.

Field-theorists do not solve the non-locality of the one-particle theory. Rather, they avoid the problem by claiming that physicists do not actually measure the projection operator onto the subspace of one-particle states in a region; instead, they measure an average

$\int \phi(\mathbf{x}, t) f(\mathbf{x}, t) \mathrm{d}^3 x \mathrm{d} t$ of a field strength over the region. If this is large, the physicist infers that there is something in the region where $f \neq 0$. Usually, the large value of the field $\phi(\mathbf{x}, t)$ in a region can be used to trigger a counter or leave a track on a photograph. This is then interpreted in terms of particles, but only to the accuracy of the resolution. Naturally, to be detectable, the particles must be described by fields which interact.

People sometimes wonder why there is no operator for time in quantum mechanics, just as there is for position. Especially in a relativistic version, they would like to write down time–energy uncertainty relations. Of course the same 'problem' occurs in classical mechanics; we can measure the position of the particle at a given time, and the position observable is just the coordinate q regarded as a function on phase-space; there is no function on phase-space giving the time a particle is at: time is merely a parameter. In classical mechanics a relativistic particle is often described by a world-line, and this goes over into a geodesic for a massless particle in general relativity. But the classical description of the interaction between world times suffers from causality problems, and there are some powerful 'no interaction' theorems. The classial description of the equations of motion of a relativistic particle in a force-field is not causal, though it is even today to be found in textbooks on special relativity. The causality problem does not afflict the Maxwell equations or Klein–Gordon equations and quite a successful classical theory exists of non-linear local field equations.

Non-relativistic quantum theory is based on the Schrödinger equation; as discovered by Bargmann, the state-space of a single free particle carries a representation of the Galilean group of space-time transformations. The three position operators then arise as the infinitesimal generators of the transformations changing from one inertial frame to another, which form a commutative set and transform as a three-vector. There is no part of the group providing a fourth commuting operator to act as time, which remains as a real parameter.

The relativistic, i.e. Lorentzian, version of the change of inertial frame is the change of Lorentz frame (the boosts). The three infinitesimal generators of boosts in the x, y, z directions (sometimes called the kick operators, K_x, K_y, and K_z) are Hermitian operators but do not commute with one another, though they do transform as a three-vector. Again, there is no operator that could do as a time-component to make a four-vector. The same applies to other position three-vectors, e.g. the Newton–Wigner position operator, which is otherwise the most satisfactory.

If this happens, for these wave equations, *ipso facto* relativistic, it suggests that we should cease trying to find a time-operator and indeed should give up the one-particle position operator as well. Indeed, particles in interaction are observed to appear and disappear with great ease. What happens to the position operator of a particle that is no longer there? And how can a state-space acquire new position operators not there before? Only the second-quantized formalism can cope with these phenomena, and at the same time treat the particles as indistinguishable. The symmetry between space and time is restored by regarding position, as well as time, as a *classical* parameter, so that the classical four-vector $(\mathbf{x}, t)$ is the space-time point at or around which we make our *field* measurement: in this way we lose the tagging of individual particles at positions and achieve a formalism allowing for, and indeed predicting, the creation and annihilation of particles.

The interacting theories such as quantum electrodynamics, Yukawa theory, and the Fermi theory of weak interactions were set up by adding an interaction term $\mathscr{L}_1$ to the free Lagrangian. This was done in analogy to what happens in classical electrodynamics, in which $\mathscr{L}_1(x) = A_\mu(x) j^\mu(x)$. In an attempt to preserve Lorentz invariance, local commutativity, and primitive causality for the interacting theory, only local polynomial functions of the fields were considered as possible candidates for $\mathscr{L}_1(x)$. The resulting non-linear theory was solved by Rayleigh–Schrödinger perturbation methods. For QED this led to a theory able to predict accurate values for Compton scattering, Møller scattering, energy levels of hydrogen, and the magnetic moments g of electrons and muons. Fermi theory gives good predictions to weak interactions if the approximation is only taken to first order.

Higher-order corrections to these theories are divergent. This is not surprising since the fields themselves are distributions, and non-linear functions of distributions are ambiguous or divergent, e.g. $(\delta(x))^2 = \delta(0)\delta(x)$. But after some adjustment, known as renormalization, the infinities of QED are removed and the resultant finite answers agree (in the Lamb shift and the $g-2$ prediction) to seven decimal places with the experimental values.

If the remaining force, nuclear force, had been as weak as these two, and had given good agreement with experiment, then the cult of rigour (called the *Feldverein* by Pauli) would probably not have arisen. It was not that Yukawa theory gave the wrong answers, but that it could not make any reliable predictions. One could remove the divergences by renormalization but, unlike in QED, one obtained a power series in large coupling constant, and 'higher corrections' got larger and larger. Non-perturbative approximate methods involved arbitrary parameters, and gave an approximate theory violating one or other of the desired principles (relativity, positivity, locality, etc.). Such models are too easy to construct, and lose predictive power— they are ten a penny. When I was a PhD student I asked my supervisor what the quantized field $\phi(\mathbf{x}, t)$, is. He said: 'It is the operator in Hilbert space assigned by the physicist to the classical nuclear field at $\mathbf{x}$ at time t according to the correspondence principle.' But he could not tell me what Hilbert space, what operator, except for the free field. Dirac sometimes wrote as if there is the physical Hilbert space *out there*, and that is the one we should use. But we cannot do this. We must specify our Hilbert space and also a correspondence between observables out there and operators. Relations between our operators then predict relations between the physical quantities they represent. To test a model, we must study its predictions. The trouble is that Yukawa theory (and also QED when one looked hard at it) is not well defined, and different approximations give quite different predictions, rather than very similar predictions. One cannot tell whether one is testing the theory or the method.

How differently we proceed in mathematics. In 1939 Wigner was very successful in making the definition: an elementary particle *is* an irreducible projective representation of the Poincaré group, $\mathscr{P}$, with mass ≥ 0 and energy ≥ 0, and spin $s \in \{0, 1/2, 1, \ldots\}$. He did not merely say that a particle is well described by such a representation: this would leave the word particle still undefined. Thus a particle *is* a pair $(\mathscr{H}, U_{[m,s]})$ where $\mathscr{H}$ is a Hilbert space, and U is a unitary continuous action of $\mathscr{P}$ on $\mathscr{H}$, obeying $U(a, \Lambda)U(b, M) = \omega U(a + \Lambda b, \Lambda M)$, where $a, b \in \mathbb{R}^4$ are space-time vectors, and Λ, M are Lorentz matrices, and where [m, s] are the mass and spin.

With this definition, we can do several things; we (i.e. Wigner) can find all possible (stable) particles. And we find them described by the Klein–Gordon equation, Dirac equation, Maxwell's equation, etc., up to unitary equivalence. Wigner does not predict why some possible particles do not occur out there e.g. massless scalar particles. But it is a big improvement over the previous *ad hoc* introduction of the wave equations.

Wightman[7] wanted to take the same attitude towards quantum field theory—to know when we have got an acceptable theory, and when not. Thus (1956), a scalar *Wightman theory* is a quadruple ($\mathcal{H}$, U, Ω, ϕ), where $\mathcal{H}$ is a Hilbert space, U is a representation of $\mathcal{P}$ with positive energy, Ω is the unique state in $\mathcal{H}$ invariant under $U(a, \Lambda)$ (the vacuum) and ϕ is an operator-valued distribution obeying (i) $U(L)\phi(x)U^{-1}(L) = \phi(Lx) = \phi(\Lambda x + a)$ (scalar covariance); (ii) $[\phi(x), \phi(y)] = 0$ if x is spacelike to y. Wightman took some, but not all, the desirable features of the free field and made them into axioms. He omitted the canonical commutation relations $[\phi(\mathbf{x}, 0), \dot\phi(\mathbf{y}, 0)] = i\delta(\mathbf{x} - \mathbf{y})$, since these are not valid in perturbation theory if there is wave-function renormalization. He also did not require primitive causality or any particular equations of motion.

Free fields obey these axioms, as do currents like $\frac{1}{2i}(\phi^*\partial_\mu\phi - \phi\partial_\mu\phi^*)$. But interacting models are hard to come by.

We can add more structure to the concept above. Thus we say a field theory contains a *particle* if $U(a, \Lambda)$ has $U_{[m,s]}$ as an isolated subrepresentation. Haag's scattering theory then says that there are ingoing and outgoing free particles asymptotically constructed from the fields. These free states define the S-matrix as the operator taking us from the free ingoing states to the free outgoing states. The fact that the existence of S can be proved from the other properties leads us to believe that we are on the right track.

The Wightman programme made great strides between 1955 and 1962, when he and I wrote a book[8] describing the theory. Great hopes were raised between 1966 and 1973, when Nelson, Glimm, and Jaffe[9] constructed $\lambda\phi_2^4$ and $\lambda\phi_3^4$. These are Wightman theories in $1 + 1$ and $2 + 1$ space-time dimensions, constructed as the limit of a cut-off approximate model with a quartic self-interaction. They also constructed Y_2, Yukawa theory, and some gauge fields in $1 + 1$ dimensions.

The limits taken to construct $\lambda\phi_4^2$ are not always unique: if λ is large,

---

[7] R. F. Streater and A. S. Wightman, *PCT Spin Statistics and All That* (New York: Benjamin 2nd edn. 1978).
[8] Streater and Wightman, *PCT Spin Statistics*.
[9] Glimm and Jaffe, *Quantum Physics*.

there are two limits[10], one with $\langle\phi\rangle_+ > 0$ and the other with $\langle\phi\rangle_- < 0$ (these are the expectations in the true ground states). Other solutions without a ground state at all can then be constructed[11] by building (in one-dimensional space, remember) a soliton between the two solutions, with the soliton state converging to $|>_-$ to the left and $|>_+$ to the right. The existence of such representations shows that neither of the vacuum theories contains all the physical particles. It is likely that the soliton is a Fermion with Yukawa interaction with the field ϕ. Thus field theory in $1+1$ and $2+1$ dimensions has a lot of good properties.

In 1972 Wilson[12] predicted that $\lambda\phi_4^4$ would if renormalized in the same way, lead only to a trivial theory. This has now (all but) been proved. Also, although Y_2 and Y_3 are non-trivial Wightman theories, Y_4 is predicted to be trivial, i.e. free or generalized free fields. This argument also applies to $(QED)_4$! These triviality statements are nowhere nearly proved, but led physicists to study QCD (quantum chromodynamics), in which Wilson's 1972 arguments do not work. We now have a puzzle: what does the Feynman–Dyson series for $\lambda\phi_4^4$ mean if there is no theory to which it is an asymptotic series? One possibility is that it is the asymptotic expansion of the Green's functions of a theory with indefinite metric; such a theory cannot be reached by the methods of Glimm and Jaffe, which always (if they converge) lead to a Wightman theory (in the case of $\lambda\phi_4^4$, a free one). Another (faint) hope is that a more subtle way of proceeding might lead to a non-trivial answer, such as the pre-asymptotic method of Symanzik[13]. Such methods are very hard. For $(QED)_4$ the hope is that it is not self-consistent unless embedded in $(QCD)_4$. The remarkable agreement of the perturbation series with experiment could then be possible even if $(QED)_4$ is trivial, if at low energy QCD has an asymptotic expansion agreeing with the QED perturbation series.

Gauge theories like QCD are very difficult to fit into Wightman's axiomatic scheme. A lot of serious axiomatization remains to be done. The trouble is that even the gluon fields $F_a^{\mu\nu}(x)$ are not gauge

[10] Ibid.
[11] J. Fröhlich, 'New Superselection Sectors ('Soliton States') in Two-dimensional Bose Quantum Field Models', *Commun. Math. Phys.* **47** (1976), pp. 269–310.
[12] K. Wilson, Lecture at the XVI Conference on High-energy Physics, Chicago and Batavia, 1972. Proc. published by NAL, Batavia (ed. J. D. Jackson and A. Roberts), Vol. 2, p. 169.
[13] K. Symanzik, *Commun. Math. Phys.* **45** (1975), p. 79.

invariant, and so will not be local or causal in general gauges. However, gauge-invariant quantities like $\text{tr}(F_a{}^{\mu\nu}Fa_{\mu\nu})$ are pointwise products of distributions, a dubious beginning to a rigorous treatment. There are gauge-invariant quantities, the Wilson loops, but these are not localized at points. Moreover like all gauge theories, to be manifestly covariant, an indefinite metric must be used. Khoruzy[14] has studied such spaces (Pontrjagin spaces, Krein spaces) from the point of view of quantum field theory; they are very complicated.

It should be possible to fit QCD into Haag's general axiomatic scheme[15] of observables. Haag does not insist on exact localizability of a field at a point; he assumes that all the observables measurable in some region of space-time are local, i.e. commute with observables spacelike separated, and transform under $\mathscr{P}$ into observables located in the transformed region. All the free fields and the solved models in 1 + 1 and 2 + 1 dimensions obey Haag's postulates as well as the Wightman axioms.

It should be mentioned that scientists must learn the sort of thing that can or cannot be postulated. The favourite theme of some authors is that, since there is only one world, the laws of physics must be such as to admit a unique answer. This kind of 'postulate' is alien to a truly mathematical approach: any list of axioms with a unique solution will admit many solutions if some of the axioms are omitted—this would not mean that the remaining axioms were incorrect. In the latest string theories, even the dimension of space-time is left open, and is supposed to be predicted by the requirements of Lorentz invariance, etc. However, since the known models in 1 + 1 and 2 + 1 space-time obey such general laws, one cannot rule out $d = 2$ or $d = 3$ as the dimension of space-time. It is $d = 4$ that gets ruled out!

In 1960 the S-matrix progamme was launched; Chew rejected field theory but took the consequential properties of the S-matrix as axioms. These properties were considered so restrictive that Chew conjectured that there would be one solution (he meant two, as $S = 1$ is obviously possible). Thus he believed that all masses and coupling constants could be predicted. In 1 + 1 or 2 + 1 dimensions this is now known to be false, even though the S-matrix equations are just as intricate as in four dimensions. It would be a strange twist if Chew turned out to be correct in $d = 4$ in the (original) form: there *is* a unique

[14] S. Khoruzy, *Commun. Math. Phys.* (forthcoming).
[15] Haag and Kastler, 'An Algebraic Appoach to Quantum Field Theory'.

solution; namely $S=1$. In that case there would be no interacting Wightman theory in $d=4$, which had particles and was asymptotically complete—we would need to start again. Perhaps the new axioms should not only include stringlike objects, but surfaces and regions as well. Albeverio and Høegh-Krohn[16] have made a start on such ideas.

[16] S. Albeverio, R. Høegh-Krohn, and H. Holden, 'Markov Cosurfaces and Gauge Fields', *Phys. Austr.* **26** (1984) pp. 211–31.

9
The Algebraic Approach to Quantum Field Theory
SIMON SAUNDERS

0. INTRODUCTION

The fundamental entity in the algebraic approach to quantum theory is an abstract C*-algebra $\mathscr{A}$, together with its set of states $\mathscr{T}$, that is, the set of all positive linear functionals on $\mathscr{A}$. I shall call such a pair an *(abstract) quantum system*. The theory of C*-algebras and their representations provides the most powerful abstract approach to quantum theory; unlike the propositional lattice approach it also has direct applications in mainstream physics, and in a certain sense is even forced upon us in quantum field theory (QFT) (see sections (2.11), (2.16)–(2.18)). Mathematically, the subject is a complex of analysis and algebra; it is a rather wild area involving high degrees of infinity. Philosophically it is most strongly associated with operationalism; in part this reflects the predilections of its originators, and in part the continuing philosophical appeal of positivism within the mathematics community. (I do not know why this is so.)

In section 1 I give a brief historical orientation to the subject and propose a sketch of what a realist interpretation might look like. In section 2 the abstract theory is further developed along with its representation theory, in application to QFT, including thermodynamic representations. The final section is devoted to the measurement problem. The aim is to give an informal survey of the theory, so proofs are omitted. In such a mathematically sophisticated area, there is a pay-off between accuracy and accessibility; but the balance in the literature is all to the former.

1. CONSTRUCTION OF AN ABSTRACT C*-ALGEBRA

1. It is well known that quantum mechanics was first developed as an algebraic theory; this is most obvious in the Born and Jordan paper of

© Simon Saunders 1987.

1925 and that of Dirac the following year, but it also underlies Heisenberg's original breakthrough of 1925. One recalls that he was thinking of how the sequences of numbers, representing the Fourier coefficients in the expansion of a dynamical variable with respect to the Bohr frequencies, could be algebraically combined in a way which preserved this interpretation.

2. If one thinks of the algebraic structure of the magnitudes associated with a physical system as fundamental, the introduction of wave mechanics was actually a complication, because a matrix algebra is much simpler than an algebra of differential operators on a function space and one has the spurious matter–wave interpretation to confuse one further. The fact that physicists found the wave theory much easier to understand is evidence that the algebraic point of view was not widely appreciated. Dirac was the great exception; most of his important work at this time—the Dirac correspondence, the second quantization formalism, and even his factoring of the Klein–Gordan equation to obtain the Dirac equation—came from this algebraic point of view. In particular his work on q-numbers led directly to Jordan's development of the idea of an abstract Jordan algebra.

3. Another inspiration of Heisenberg's work was the renunciation of semi-classical models for the atom; in particular he worked directly with Fourier series in the Bohr frequencies rather than with frequencies associated with the periodic motion of electrons within the atom. No data could be obtained on the latter frequencies, whereas the Bohr frequencies were directly associated with spectroscopic data; hence the dogma: matrix mechanics is concerned only with measurable quantities, or 'observables'.

4. Faced with the burgeoning analytic structure of wave mechanics and the daunting problems of its generalizations to a relativistic theory, Jordan sought once more to free the theory from 'unphysical' elements and work directly with the algebra of observables.

To develop this idea, he appealed to the limitations of experiment; on the interpretation that an element of this algebra should correspond to what is directly observed, what possible interpretation could be placed on the *multiplication* of 'observables'? The sum and power-raising operations, and multiplication by real scalars, he considered, however, unobjectionable, and with these one can define a commutative produce ● as $A \bullet B = \frac{1}{2}[(A+B)^2 - A^2 - B^2]$. The resulting (not-necessarily distributive or associative) algebra he called an *r-number algebra* (Jordan 1933). In his later collaboration with von Neumann and Wigner some really substantial results were obtained. With two further assumptions (that if $A^2 + B^2 + \ldots = O$ then $A = B = \ldots = O$, and that $A^m A^n = A^{m+n}$), together with a condition roughly equivalent to distributivity, they were able (Jordan et al. 1934) to develop a complete classification and representation theory for the finite case, where the algebra can be generated from a finite number of elements. This algebra is now called a *commutative real Jordan algebra* (we shall say simply a Jordan algebra).

5. This methodology is characteristic of the abstract approach to quantum physics (both in algebraic and propositional lattice theories). First, one tries to find an abstract mathematical structure, and this is very speculative, because one gives up the age-old assumptions as to how one represents physical properties—they are no longer parameterized by numbers. But, second, one chooses a structure which will have a simple interpretation in terms of measurement, one tries to constrain it in this way. The paradigm in both cases is matrix mechanics.

6. A Jordan algebra $\mathcal{U}$ has a close relationship to an (abstract) C*-algebra. Let $\mathcal{V}$ be an associative (not necessarily commutative) algebra, with the symmetrized product $A \cdot B = \frac{1}{2}(AB + BA)$. A Jordan algebra isomorphic to $\mathcal{V}$, such that $A \bullet B$ is mapped to the element $A \cdot B$, is called *special*. In the finite case Jordan et al. proved that every Jordan algebra is a direct sum of special Jordan algebras, with a single exception: the algebra of all hermitian matrices of order 3 over the Cayley numbers (see Albert 1934; incidentally, the possible physical relevance of this exceptional Jordan algebra is still unknown). For special Jordan algebras, if $\mathcal{V}$ is commutative than $\mathcal{U}$ is associative. Quite generally, the self-adjoint part of any C*-algebra is a special Jordan algebra.

7. Von Neumann later tackled the infinite-dimensional case (von Neumann 1936); to obtain more control over the algebraic structure, he sought to introduce some topological tools and in particular to imitate the weak operator topology, in connection with his work at that time on rings of operators in collaboration with Murray. It was this paper which directly motivated Segal some ten years later, who was studying the representation theory of operator algebras. This paper (Segal 1947) was of fundamental importance: both because it was able to introduce a powerful new representation theory (a generalization of one due to Gelfand and Naimark 1943), and because he was able significantly to extend the conceptual basis of the

theory. Both developments came about through introducing and exploiting the notion of a *state* on the algebra.

8. This innovation was not in fact contrary to a purely algebraic point of view. Segal was actually able to prove that there *must* exist pure states, as positive linear functionals on the algebra, normalized to unity, which cannot be given as a convex sum of other states. He also showed that there are 'enough' of them to completely characterize the algebraic structure, in terms of the real numbers which they assign to the elements of the algebra (see (1.10), (1.11) below). The other major innovation of this paper—his postulate that the algebra has a *norm topology*—was also introduced in a purely algebraic way (although it takes a particularly simple form in terms of the states). In this way the notion of state did not play an *axiomatic* role in his theory.

9. It would be a lenghty task to review the theory of these abstract algebras and their subsequent variants: JBW-, W*-, von Neumann, and Σ^*-algebras (to name the most important). Taken together with the theory of their associated lattice structures (and more generally in their relationships with quantum logic) it is a vast field. But all are C*-algebras, and all have self-adjoint parts which are Segal algebras. And with few exceptions, their abstract structures have been motivated (and interpreted) either on mathematical or operationalist grounds. It is a tribute to the fertility of positivist philosophy that this should even be possible. But for all that I am not a positivist, and one has yet to see realism displaced as the driving force in theoretical physics, let alone in the laboratory. This being so, it is worth trying to interpret the structure of a C*-algebra in a realist way. It is not just that this is the only way of constructing ideas about an unobserved world, but that only then can one give an account of the process of observation as a special kind of physical process. If the basic entities in a theory refer to operational procedures, it is just incoherent to use that theory to produce detailed models of those same procedures.

10. A realist account (with conventional overtones) might go something like this: starting from the intuitions of space-time, one has a space which contains a class of physical entities (in correspondence with, say, the field of all open bounded subsets of a Euclidean space) some of which may be completely characterized by the assignment of a real number (we shall then speak of *value assignments*). Those which cannot be fixed in this way nevertheless always have some (real) *measure*. As time evolves this polarization changes; those entities

The Algebraic Approach to Quantum Field Theory 153

which can *always* be completely characterized by value assignments we shall think of as classical observables. In general their value assignments vary in a stochastic way. In certain circumstances (measurement processes) the statistical behaviour of a classical observable may yield data on the measures or value assignments of other entities (quite possibly in a sub-volume of the classical object which it describes). The last fundamental assumption is that this class of physical entities has an algebraic structure, extended from the algebraic structure of those observables which have values assigned to them. In particular the algebra is a linear space over the real numbers, distributive with respect to multiplication. We can now make sense of what we mean by a (numerical) complete characterization: the algebraic operations are respected by the value assignments. More generally, the *linear space structure* is always preserved by a measure assignment.

11. The mathematical expression of these ideas is as follows: consider an abstract algebra $\mathcal{U}$ and a class of maps $\mathcal{T} \ni \phi : \mathcal{U} \to \mathbb{R}$ with values $\langle \phi; A \rangle$, $A \in \mathcal{U}$. For each $\phi \in \mathcal{T}$, let $\mathcal{U}_\phi$ be a subset of $\mathcal{U}$ which is 'completely characterized' by numerical assignments; by this we mean that the usual algebraic structure of the set of complex numbers $\langle \phi; \mathcal{U}_\phi \rangle$ also describes the algebraic structure of the set $\mathcal{U}_\phi$. That is, for any $A, B \in \mathcal{U}_\phi$ and any $\lambda \in \mathbb{R}$ and integer n there exists an element in $\mathcal{U}_\phi$ (denote $A + \lambda B$) such that $\langle \phi; A + \lambda B \rangle = \langle \phi; A \rangle + \lambda \langle \phi; B \rangle$, and an element in $\mathcal{U}_\phi$ (denote AB) such that $\langle \phi; AB \rangle = \langle \phi; A \rangle \langle \phi; B \rangle$; in particular there exists A^n such that $\langle \phi; A^n \rangle = \langle \phi; A \rangle^n$. Suppose now that for every A in $\mathcal{U}$ there is a ϕ such that $A \in \mathcal{U}_\phi$; in this way we extend the algebraic structure to the whole of $\mathcal{U}$. We also suppose that each $\mathcal{U}_\phi$ has a unit and that for every ϕ the set $\mathcal{U}_\phi$ is non-empty (perhaps only the unit). It then follows that the mappings $\mathcal{F}$ are actually positive linear functions on $\mathcal{U}$, normalized to unity (because $\langle \phi; I^2 \rangle = \langle \phi; I \rangle^2 = \langle \phi; I \rangle$).

12. In the usual terminology $\mathcal{T}$ is the set of *states*, the $\langle \phi; A \rangle$ is called the *expectation value* of A, and the elements of $\mathcal{U}$ are called *observables*. We shall use this terminology from now on, although the connotations are unfortunate (in our account only classical observables are *directly* observable, and the probabilistic overtones of 'expectation value' are only appropriate when an observable has measure equal to the mean value of some classical parameter).

13. Let $\mathcal{T}_A$ be that subset of $\mathcal{T}$ such that $\langle \phi; A^n \rangle = \langle \phi; A \rangle^n$ for all $\phi \in \mathcal{T}_A$; this subset is the set of dispersion-free states of A. Because for

every $A \in \mathcal{U}$ there is a ϕ such that $\mathcal{A}_\phi \ni A$, the set $\mathcal{T}_A$ is non-empty. Obviously the numbers $\langle \phi; A \rangle$, $\phi \in \mathcal{T}_A$ are the possible value assignments that can be made of A. Henceforward, and as a matter of mathematical convenience, we make two assumptions; that *every* observable has a *finite* maximum value (we assume $\mathcal{U}$ is an algebra of *bounded* observables), and that any sequence A_i which is Cauchy with respect to the measure given by any state in $\mathcal{T}$ and which converges to a finite value exists as an element of $\mathcal{U}$; that is $\mathcal{U}$ is a metric space with norm $\|A\| = \sup_{\phi \in \tau_A} (\phi; A^2)^{1/2}$, and that $\mathcal{U}$ is *complete*. This step is analogous to completing the rational numbers to obtain the reals. With this structure we suppose we can give an exhaustive description of the world; we therefore make the natural assumption that there is no redundancy, i.e. that for every $A, B \in \mathcal{U}$ there is a state $\phi = \mathcal{T}$ such that $\langle \phi, A \rangle \neq \langle \phi, B \rangle$, and that for every $\phi, \psi \in \mathcal{T}$ there is an observable $A \in \mathcal{U}$ such that $\langle \phi, A \rangle \neq \langle \psi, A \rangle$. The algebra $\mathcal{U}$ constructed in this way (the definition of the sets A_ϕ must be made a little sharper) satisfies all the postulates of Segal (for a similar reconstruction see Emch (1972); note that our assumption of distributivity replaces his structure axiom 6). It is therefore a *Segal algebra*.

14. There is one further assumption, which is actually crucial, but which has no obvious physical interpretation: that is that the algebra is *special*. The Segal algebra $\mathcal{U}$ must then be a sub-algebra (with respect to the commutative product) of the embedding algebra $\mathcal{V}$. How to recover it from $\mathcal{V}$? The easiest way is to have a map* on $\mathcal{V}$ under which $\mathcal{U}$ is mapped on to itself; thinking of $\mathcal{U}$ as the 'real part' of the 'complex' algebra $\mathcal{V}$ one is led to consider * as an *involutive antiautomorphism* (so that $A^{**} = A$, $(AB)^* = B^*A^*$). One might also hope to connect the existence of such a map with the existence of an *orthocomplement* on the lattice of idempotents in $\mathcal{U}$ (cf. Kakatuni and Mackey 1944). Elements $A \in \mathcal{V}$ such that $A^* = A$ are called *self-adjoint*. We henceforward assume that the space $\mathcal{V}$, equipped with such a map, exists, with self-adjoint part isomorphic to $\mathcal{U}$. We also make the natural assumptions that * is norm-preserving, and that $\|A\|^2 = \|A^*A\|$ for all $A \in \mathcal{V}$.

15. If $\mathcal{V}$ is a *real* vector space, it is called an *R*-algebra*. If it is *complex*, a *C*-algebra*. The GNS construction also exists for R*-algebras yielding a concrete representation on a *real* Hilbert space. Unlike the Piron representation theorem in quantum logic (see, e.g. Varadarajan 1968) and the representation theory of r-number

algebras (Jordan et al. 1934), where one obtains a Hilbert space over one of the reals, the complex numbers, or the quaternions, the division ring which enters into the Hilbert space realization via the GNS construction is always that of the abstract algebra $\mathcal{V}$. It will take us too far afield to motivate the assumption that this division ring is, in fact, the complex numbers; this assumption is more properly made at the level of applications of the theory, if one is quantizing a classical theory (cf. Segal 1967), or following a development of the representation theory in the real case (see, e.g., Stueckelberg et al. (1962)). Nevertheless, because it takes us directly to the familiar formalism of quantum theory, we make the assumption that the division ring is in fact the complex numbers $\mathbb{C}$ at the present stage. We use the symbol $\mathcal{A}$ to denote the abstract C*-algebra in which we assume $\mathcal{U}$ can be embedded; note that every element in $\mathcal{A}$ can be written as the sum (with complex coefficients) of self-adjoint elements in $\mathcal{U}$; correspondingly the properties of $\mathcal{A}$, the action of states in $\mathcal{T}$, and so on, are fixed by those of $\mathcal{U}$ by a straightforward transport of structure.

16. We have carefully abstained from placing an interpretation on an observable which is assigned a measure at some instant; the difficulty here is not to find an interpretation, but rather to make it do work: stochastic fluctuation, state of potentiality, propensity, indeterminacy, fuzziness, lack of definition—there are many concepts at our disposal. They are either too vague, or lead to unworkable systems (contextual hidden variable theories, for example). The measurement theory developed in section 3 gives us some guide-lines; but agnosticism on this score is not tantamount to the abandonment of realism.

2. REPRESENTATION THEORY AND APPLICATIONS IN QUANTUM FIELD THEORY

1. We begin with a fundamental result of Segal that is true of any Segal algebra (and a fortiori of the self-adjoint part of a C*-algebra).

Theorem: every associative Segal algebra $\mathcal{U}$ is isomorphic (algebraically and metrically) with the algebra C(X) of all real-valued, continuous functions A on a compact Hausdorff space X, where the addition, scalar multiplication, and powers of a function are defined in the usual way and where the norm is defined by $\|A\| = \sup_{x \in X} |A(x)|$. Furthermore, every state ϕ in $\mathcal{T}$ has the form $\langle \phi; A \rangle = \int_X A(x) d\mu(x)$ where μ is a regular Borel measure on X such that $\mu(X) = 1$. Conversely, every such measure generates through the above integral a state ϕ on $\mathcal{U}$. (For proof see Segal (1947); recall that a topological space X is *Hausdorff* is for any x, $y \in X$, x and y have disjoint neighbourhoods, that its *Borel sets* are the elements of the σ-ring generated by the family of all closed subsets of X, that a *Borel measure* is a finite σ-additive measure on the Borel sets which is *regular* if for any Borel set E $\mu(E) = $ l. u. b. $\{\mu(S) \ S \subset E, \ S \ \text{closed}\}$.)

When $\mathcal{U}$ is embedded in a C*-algebra $\mathcal{A}$ associativity with respect to the symmetrized product is equivalent to commutativity of the product in $\mathcal{A}$; therefore a commutative quantum system has self-adjoint part isomorphic to the canonical representation of classical mechanics (for bounded observables, that is on a compact phase space). The space X is actually constructed as the set of states on this algebra, with the isomorphism $\chi: \mathcal{U} \to C(X)$ given by $\chi(A)(\rho) = \langle \rho; A \rangle$. The pure states are defined in the usual way, as the extreme points of the space of states $\mathcal{T}$; one can show that these are dispersion-free, in the sense that $\langle \rho; A \rangle^2 - \langle \rho; A^2 \rangle = 0 \ \forall A \in \mathcal{U}$ when ρ is pure. The pure states are then the points of the set X.

This theorem also gives us the spectral theorem, because any observable generates a commutative sub-algebra (nowadays it provides the standard proof of the spectral theorem; recall also the familiar result that the spectrum of a bounded operator is always compact).

2. Important though this theorem is, it is not typical of the representation theory of C*-algebras. More representative in the commutative case is a *Koopman System*: let $L^2(X, \mu)$ be the Hilbert space $\mathcal{H}$ of all functions $\psi: x \to \mathbb{C}$, which are square integrable with respect to the measure μ associated with the given state ϕ by the above theorem. For every A in $\mathcal{U}$ define the bounded self-adjoint operator $\pi_\phi(A): \mathcal{H}_\phi \to \mathcal{H}_\phi$ by $(\pi_\phi(A)\psi)(x) = A(x)\psi(x)$ for all ψ in $\mathcal{H}_\phi$. Then π_ϕ is a *representation* of $\mathcal{U}$ (i.e. a map from $\mathcal{U}$ into $\mathcal{B}(\mathcal{H})$, the set of all bounded operators on a Hilbert space, which preserves the algebraic structure). Further, the vector Φ in $\mathcal{H}_\phi$ defined by $\Phi(x) = 1$ for all $x \in X$ satisfies

(i) $\langle \phi; A \rangle = (\Phi, \pi_\phi(A)\Phi)_{\mathcal{H}_\phi}$

(ii) the linear manifold $\{\pi_\phi(A)\Phi | A \in \mathcal{U}\}$ is dense in $\mathcal{H}_\phi$, i.e. Φ is *cyclic*.

The Koopman systems are instructive, because they illustrate in the classical case the general features of the representation theory of C*-algebras. Most important of these is the *dependence of a representation* on the state of the system. If the state ϕ is changed, we change the measure with respect to which the L^2 space is defined. If the measures have different null sets, the observables will have different norms in the corresponding representations and the kernel of the representations will differ (the set of observables which are mapped onto zero). The other important features concern the properties of the unit vector Φ; (i) tells us that it can be identified with the state that generates the representation, whilst from (ii) we learn that we can get as close as we like to any vector in the space of states associated with this representation by transforming Φ under $\pi_\phi(\mathcal{U})$. This shows just how dependent the representation is on the generating state ϕ, because from ϕ alone one can obtain all the information on $\mathcal{U}$ that can be obtained from *any* state in the state space that it generates; suppose there is a state ψ in $\mathcal{H}_\phi$; since Φ is cyclic, this state can be approximated as closely as desired by $\pi_\phi(A)\Phi$ for some $A \in \mathcal{U}$. But then, for any $B \in \mathcal{U}$ one has $(\psi, \pi_\phi(B)\psi) \simeq (\pi_\phi(A)\Phi, \pi_\phi(B)\pi_\phi(A)\Phi) = (\Phi, \pi_\phi(A)^*\pi_\phi(B)\pi_\phi(A)\Phi) = \langle \phi; A^*BA \rangle$; in this way one can replace the description of the system by the set of numbers $\langle \psi; \mathcal{U} \rangle$ by the set $\langle \phi; A^*\mathcal{U}A \rangle$.

3. When we come on to the general case (a non-associative Segal algebra) we only have a representation theory when $\mathcal{U}$ is embedded in a C*-algebra $\mathcal{A}$.

Theorem: given an abstract quantum system $\mathcal{A}, \mathcal{T}$, for each state $\phi \in \mathcal{T}$ there exists a canonical representation π_ϕ of $\mathcal{A}$ as a (not necessarily commutative) algebra of bounded operators acting on a Hilbert space $\mathcal{H}_\phi$, with cyclic vector Φ such that: $\langle \phi; A \rangle = (\Phi, \pi_\phi(A)\Phi)_{\mathcal{H}_\phi}$ for all $A \in \mathcal{A}$. *Every* representation can be obtained in this way. (Proof: Gelfand and Naimark (1943); Segal (1947).)

In this construction (*the GNS construction*) the Hilbert space $\mathcal{H}_\phi$ is actually defined as a quotient algebra of $\mathcal{A}$; loosely speaking, one defines equivalence classes $\chi(A)$ such that all observables in $\chi(A)$ differ by an observable which lies in the algebraic subspace of $\mathcal{A}$ which ϕ maps onto zero. Each such equivalence class $\chi(A)$ fixes a vector in a pre-Hilbert space with inner product $(\chi(A), \chi(B)) = \langle \phi; A^*B \rangle$ (non-degenerate and positive-definite by construction). Completing this space then gives us a Hilbert space, the strong closure of the vectors $\chi(A)$; each element A in $\mathcal{A}$ is then represented by that element in $\mathcal{B}(\mathcal{H}_\phi)$, which acting on a vector $\chi(B)$ yields the vector $\chi(AB)$. It is obvious that $\chi(I)$ is cyclic, because $\pi_\phi(\mathcal{A})\chi(I) = \chi(\mathcal{A}I) \sim \chi(\mathcal{A})$ which is dense in $\mathcal{H}_\phi$ by construction.

4. To proceed further we need some more definitions. A representation π is called *faithful* if the kernal of π is zero; *irreducible* if there is no non-zero subspace of $\mathcal{H}_\pi$ stable under $\pi(\mathcal{A})$ (i.e. no subspace S such that $\pi(\mathcal{A})S \subset S$). It is clearly helpful to distinguish states which are given as vector states of a representation from the abstract states in $\mathcal{T}$; accordingly let B_π be vectors in $\mathcal{H}_\pi$, $\mathrm{co}(B_\pi)$ the convex linear hull of B_π (the set of all density matrices in a representation), and note that every state $\rho \in \mathrm{co}(B_\pi)$ determines a state in $\mathcal{T}$, denote $j_\pi(\rho)$, by $\langle j_\pi(\rho); A \rangle = \mathrm{Tr}[\rho\pi(A)]$ for all $A \in \mathcal{A}$ ($j_\pi(\rho)$ clearly vanishes on ker π). We denote the set of all states in $\mathcal{T}$ which can be given as vector states or density matrices in some representation π (i.e. $j_\pi(\mathrm{co}(B_\pi))$) as $\mathcal{T}_{j(\pi)}$; this is not the same thing as the set of all states on $\pi(\mathcal{A})$, which is itself a concrete C*-algebra; we denote this set of states by $\mathcal{T}_\pi$.

5. If we consider again the Koopman systems, it is clear that there is no hope that two representations are unitarily equivalent if the states which generate them do not determine quasi-equivalent measures (i.e. with the same null-sets). One might think this is because the representations are not *faithful*. Suppose we have a faithful representation π; then clearly $\mathcal{T}_\pi = \mathcal{T}$. Intuitively, one might also think that $\mathcal{T}_{j(\pi)} = \mathcal{T}$, but the answer is negative: the density matrices even in a faithful representation do not give us all the states in $\mathcal{T}$. Correspondingly, when we have two representations which determine distinct subsets of $\mathcal{T}$, i.e. $\mathcal{T}_{j(\pi_1)} \neq \mathcal{T}_{j(\pi_2)}$ then the representations are *unitarily inequivalent*. An equivalent criterion of unitary equivalence can be given in terms of the vector states of the representations, i.e. $j_\pi(B_\pi)$; since the generating state of a representation is always cyclic, unitary equivalence of two representations $\pi_{\phi_1}, \pi_{\phi_2}$ is equivalent to the condition that $\phi_1 \in j(B_{\phi_2})$ and $\phi_2 \in j(B_{\phi_1})$. We now note that ϕ is pure if and only if π_ϕ is irreducible (Emch 1972, Ch. 3, 87); clearly then *every* vector in $\mathcal{H}_\phi$ is cyclic, so that for irreducible representations unitary equivalence is assured if *some* vector state in $J(B_{\phi_1})$ is contained in $j(B_{\phi_2})$ and vice versa. To see the sorts of possibilities that can arise, the number of irreducible unitarily inequivalent representations of $\mathcal{A}$ when $\mathcal{A}$ is $\mathcal{B}(\mathcal{H})$ (with $\mathcal{H}$ separable) is 2^c, where c is the cardinality of the continuum, each on a non-separable Hilbert space (Kadison 1967).

6. This fact raises a host of questions. How do we interpret these unitarily inequivalent representations? Can we find physical criteria which pick out a particular representation as the 'true' one? Is unitary

equivalence too strong, and can we find some other kind of equivalence instead? Is an abstract quantum system too general a structure, and should this approach be abandoned? If not, should we rather abandon the idea of obtaining a concrete formalism in dynamical physics? What are these states anyway that are not obtained in the conventional formalism, and what has happened to Gleason's theorem? From now on we devote ourselves to these questions, roughly in the reverse order. We start with Gleason's theorem.

7. Gleason's theorem plays an important role in the quantum logic approach and stands at the heart of the Kochen–Specker 'paradox' (Gleason 1957; Kochen and Specker 1967); it states that every probability measure on the lattice of subspaces of a Hilbert space or dimension greater than two is given as a density matrix.

The fundamental reason why this theorem does not apply to the states of a C*-algebra is that the definition of state is not the same as the definition of a probability measure; the latter is σ-*additive*, i.e. for every countable family of mutualy orthognal projections P_i, a probability measure ρ satisfies $\rho(\Sigma_i P_i) = \Sigma_i \rho(P_i)$. This continuity condition cannot be applied directly to an arbitrary C*-algebra; *the idempotents in $\mathscr{A}$ do not necessarily form an orthocomplete lattice* (that is, $\Sigma_i P_i$ does not in general exist in $\mathscr{A}$). This is the fundamental feature of the algebraic approach which distinguishes it from quantum logic (or lattice) axiomatizations of quantum theory. The various abstract theories which have been proposed to subsume both the algebraic and lattice theories (for example Plymen (1968a and b) all enlarge the algebra so that it is orthocomplete.

I shall merely describe the situation for one class of C*-algebras which contain 'enough' projections to formulate the conditions of Gleason's theorem, in particular such that the complete lattice of subspaces $\mathscr{L}(\mathscr{H}_\pi)$ lies in $\pi(\mathscr{A})$; that is when $\pi(\mathscr{A})$ is an (irreducible) *von Neumann algebra*. Given a concrete C*-algebra $\mathscr{A}$ as a subalgebra of $\mathscr{B}(\mathscr{H})$, for some fixed Hilbert space $\mathscr{H}$, the *commutant* $\mathscr{A}'$ is the subset of $\mathscr{B}(\mathscr{H})$ which commutes with all members of $\mathscr{A}$. If $\mathscr{A}$ equals the *bicommutant* $\mathscr{A}''$ then $\mathscr{A}$ is a von Neumann algebra. Such an algebra has the nice property that it is generated by its projections; the bicommutant of the set of projections in $\mathscr{A}$ gives the whole algebra, as does also the closure of this set with respect to a number of topologies, all weaker than the norm topology.

Now, suppose $\mathscr{A}$ is irreducible and von Neumann; is then the definition of a state equivalent to the definition of a probability measure? A state, as a positive linear functional on $\pi(\mathscr{A})$, is continuous in the *norm* topology; we now have the following theorem (Dixmier 1957, ch. 1, sec. 3): a state ϕ in $\mathscr{T}_\pi$ lies in $\mathscr{T}_{j(\pi)}$ if and only if ϕ is continuous on $\pi(\mathscr{A})$ with respect to either the ultraweak or the ultrastrong topology (that is a state is given as a density matrix if and only if this condition holds). One suspects, from Gleason's theorem, that ultraweak or ultrastrong continuity is equivalent to σ-additivity; actually it is equivalent to *total* additivity (i.e. that $\rho(\Sigma_i P_i) = \Sigma_i \rho(P_i)$ for *every* family of mutually orthogonal projectors (Emch 1972, p. 118)). We now notice that Gleason assumes $\mathscr{H}$ is separable, in which case (if $\pi(\mathscr{A})$ is to include $\mathscr{L}(\mathscr{H})$) $\pi(\mathscr{A})$ must be σ-finite (Takesaki 1979, proposition 3.19), hence admits at most countably many orthogonal projections.

The conclusion of this discussion is that, in the present theory, Gleason's premises, in so far as they can be given a meaning, are equivalent to either ultrastrong or ultraweak continuity of the states. Both of these topologies are weaker than the norm topology (so that, roughly speaking, one demands convergence on sequences of observables which are not Cauchy with respect to the norm topology).

8. Given an arbitrary von Neumann algebra (which is also of course a C*-algebra) the states which are ultrastrongly or ultraweakly continuous are called *normal*. Can one impose this condition on an abstract quantum system and thus restrict the class of states? The answer is negative: a C*-algebra is not in general closed in either of these topologies (unlike a von Neumann algebra). Intuitively, in order to reduce the number of states on the system, one has to increase the size of the algebra. One can, however, always associate a von Neumann algebra with any abstract quantum system. One does this by forming the *universal representation* of the C*-algebra (by taking the direct sum of every GNS representation for every state in $\mathscr{H}$, and thus form the concrete algebra $\oplus_{\phi \in \mathscr{F}} \pi_\phi(\mathscr{A})$ on the space $\oplus_{\phi \in \mathscr{F}} \mathscr{H}_\phi$; this representation although highly reducible, is always faithful. One then takes the bicommutant of this to obtain the universal enveloping von Neumann algebra. But one then finds that the normal states on this algebra are precisely the set of states $\mathscr{F}$ of the abstract quantum system. This fact is very useful, for in this way one reduces the study of the set of all states on a C*-algebra to the study of the normal states on its universal enveloping von Neumann algebra, but one does not thereby reduce the number of states.

The Algebraic Approach to Quantum Field Theory

9. An example of some 'unusual' states in $\mathcal{T}$: let $\mathcal{A} = \mathcal{B}(\mathcal{H})$; then there are observables in $\mathcal{A}$ which have a continuous part to their spectrum. But to *every* observable of a quantum system there exists a pure state which is dispersion-free and has an expectation value equal to any point in its spectrum. Therefore such a state cannot be given as a vector state in $\mathcal{H}$, for it would be an eigenstate contradicting the premiss that the observable has a continuous part to its spectrum. In particular one could find dispersion-free states on any bounded function of the position or momentum operators. This does not mean that such a state can never be given as a vector state in some representation; on the contrary, *every* state $\phi \in \mathcal{T}$ is a vector state in $\mathcal{H}_\phi$.

10. At this stage one might wonder how we have managed for so long with ordinary quantum mechanics. Let me quickly discuss what happens in this case. In this theory (NRQM) one has the CCRs (canonical commutation relationships):

$$[p_i, q_j] = i\hbar \delta_{ij}, \quad [p_i, p_j] = [q_i, q_j] = 0,$$

and one has the familiar Schrödinger representation of the p's and q's as essentially self-adjoint operators on an L^2 space:

$$(p_i f)(x) = -i\hbar \frac{\partial}{\partial x_i} f(x), (q_i f)(x) = x_i f(x), f \in L^2(R^n, dx^n).$$

The question is: is this representation unique, up to unitary equivalence? Actually in this form we do not have any uniqueness, unless we make some additional assumptions as to the existence of a common domain for these operators, and that is because they are unbounded. For the same reason we do not even have an algebra. But if we define the operators (setting $\hbar = 1$)

$$U(\mathbf{a}) = e^{ip_i a_i} \quad V(\mathbf{b}) = e^{iq_i b_i}, \quad \mathbf{a}, \mathbf{b} \in R^n,$$

then from the CCR's it follows that $U(\mathbf{a})V(\mathbf{b}) = V(\mathbf{b})U(\mathbf{a})e^{a_i b_i}$, as may easily be seen by differentiating this relationship with respect to a_i and then b_j, and evaluating at $a_i = b_j = 0$. Acting on f in L^2 we also have:

$$(U(\mathbf{a})f)(\mathbf{x}) = f(\mathbf{x} - \mathbf{a}), \quad (V(\mathbf{b})f)(\mathbf{x}) = e^{ib_i x_i} f(\mathbf{x}).$$

We can now turn the problem around: starting from the algebra of the U's and V's, which is indeed a C*-algebra, is there a unique representation (up to unitary equivalence) as operators on a Hilbert space? The answer is yes—if we require that they are weakly-continuous groups as maps $R^n \to \mathcal{B}(\mathcal{H})$. This last provision is essential; it is exactly the kind of condition which excludes certain states in $\mathcal{T}$ as 'non-physical'; here it is connected with the requirement that one can define self-adjoint operators which generate these unitary groups via Stone's theorem, which demands just this continuity property. This theorem is due to von Neumann; the algebra is called the Weyl algebra of the CCRs, and one proves that any faithful representation is unitarily equivalent to a direct sum of Schrödinger representations as defined above. This theorem is the reason why one can ignore all the complications of a correct choice of representation in NRQM.

11. The fundamental obstacle to a concrete approach to QFT is the failure of the von Neumann uniqueness theorem for such systems. For convenience, define the quantities $W(\mathbf{a},\mathbf{b}) = U(\mathbf{a})V(\mathbf{b})e^{-\frac{1}{2}ia_ib_i}$, so that $W(\mathbf{a},0) = U(\mathbf{a})$, $W(0,\mathbf{b}) = V(\mathbf{b})$, and $W(\mathbf{a},\mathbf{b})\,W(\mathbf{c},\mathbf{d}) = W(\mathbf{a}+\mathbf{c}, \mathbf{b}+\mathbf{d})e^{i(a_i - b_i c_i)/2}$, or if α is the pair $(\mathbf{a}, \mathbf{b})$ (a point in phase space with symplectic form w) simply: $W(\alpha)W(\beta) = W(\alpha + \beta)\,e^{iw(\alpha,\beta)/2}$. In this way one can obtain a fully covariant form of the Weyl algebra. (I shall not bother with the details.) Now in a field theory one has the formal equal time CCRs $[\phi(\mathbf{x}),\pi(\mathbf{x}')] = i\hbar\delta^3(\mathbf{x} - \mathbf{x}')$ (for a Boson scalar field ϕ with canonically conjugate field $\pi = \partial\phi/\partial t$). These are not even unbounded operators: to make sense of them we introduce the 'smeared' fields $\phi(f) = \int\phi(\mathbf{x})f(\mathbf{x})d^3x$ with f in some suitable function space S with inner product (,). One then has $[\phi(f,\pi(g)] = i\hbar(f,g)$. Exactly as before one can define the Us and Vs, with the function space S replacing phase space, or going directly to the Ws one obtains the algebra: $W(f,g)W(f',g') = W(f+f',g+g')e^{i((f,g')-(g,f'))/2}$. A state on this algebra is then fixed by giving the values of the functional $\hat{\phi}(f,g) = \langle\phi; W(f,g)\rangle$ for all $f,g \in S$.

What of the uniqueness theorem? The most general way of understanding this theorem is through the imprimitivity theory of Mackey (Mackey 1952, 1968); in the previous case one wants to find a representation of the additive group, with multipliers $e^{ia_ib_i}$, on a finite dimensional space, and Mackey's fundamental result is that these representations are in a one–one correspondence with the cohomology classes of the group, up to unitary equivalence. Because the additive group is abelian, we get uniqueness. When we go to a function space, however, we have an infinite dimensional space which is not locally compact, and the whole Mackey theory fails because it can be set up only for representations of groups on locally compact spaces.

12. Therefore in contrast to the situation in NRQM, where a general continuity requirement is enough to fix an equivalence class of 'physical' representations, we must look elsewhere in QFT. An obvious strategy is to fix on some particular state which is in some way privileged, which must occur in any 'physical' representation. From (2.5) we can expect to obtain in this way an equivalence class of representations, say all those which contain this distinguished state (if we confine ourselves to irreducible representations that will be enough to establish unitary equivalence). The standard choice since the inception of QFT has been a *vacuum* state, which satisfies the conditions of invariance under the symmetry group of the theory and which is mapped onto zero by the annihilation operator. These are enough to determine the functional $\hat{\phi}_F(f,g) = e^{-\frac{1}{2}((f,f)+(g,g))}$ $= \langle \phi_F; W(f,g) \rangle$ and thereby the state ϕ_F. The GNS representation π_{ϕ_F} is then a *Fock representation*. One might therefore say that what is typical or physical reality, what we must be able to define in the concrete theory, is a vacuum state, and this excludes all the abstract states on the abstract algebra which cannot be generated from this state in the way described in (2.2).

13. Unfortunately there is a major flaw in this strategy. The Fock representation is not fixed unless one specifies the norm on the test-function space and with different norms, one gets unitarily inequivalent representations. Neither are there obvious physical grounds for fixing this norm. One might think it must be specified even to define the algebra, but this is not true: the algebra only requires that the symplectic form $(f,g')-(g,f')$ be specified (for which there *are* physical criteria), and this can be fixed with many different choices of the inner product. There are also good reasons for believing that this problem underlies some of the pathologies of RQFT, in particular some of the divergences that occur in perturbation theory. The first indications that this might be so came from the work of Friedrichs and van Hove in the early 1950s. The latter in particular produced an exactly solvable model for the meson field in an external classical field, and showed that as the external source contracts to a point, the renormalized vacuum cannot exist in the Fock space of the bare mesons and the renormalized Hamiltonian cannot exist as an operator in the Fock space of the free theory. A general theorem was proved in 1955—due to Rudolf Haag—which confirmed that this situation is typical: the free field vacuum and the physical vacuum cannot exist as states in the same representation and a field which is

asymptotically free, i.e. unitarily equivalent to the free field as $t \to \pm \infty$, is unitarily equivalent to the free field at all times, so that it has a trivial S-matrix.

14. The view that these technical difficulties should be avoided by dispensing with concrete representations altogether (or postponing their introduction until a late stage of the theory)—that no significant physical content attaches to the choice of representation—was widespread in the early 1960s. This view was encouraged by the introduction of the idea of *weak equivalence* of representations (also called *physical equivalence* (Fell 1960). Consider a faithful representation π; then $\pi(\mathscr{A})$, as a C*-algebra, is identical to $\mathscr{A}$, and its set of states $\mathscr{T}_\pi$ is identical to $\mathscr{T}$. The difficulty is that the states which can be given as density matrices (the closed convex hull of the vector states in $\mathscr{H}_\pi$), i.e. $\mathscr{T}_{j(\pi)}$ is a proper subset of $\mathscr{T}_\pi$. Fell noticed that, nevertheless, one can approximate every state in $\mathscr{T}_\pi$ by a *net* (not a sequence) of states in $\mathscr{T}_{j(\pi)}$. The concept of nets is a rather technical one and we shall not bother with its definition here; the basic result is that the approximaton is *pointwise*, i.e. for any *fixed and finite* subset of observables in $\mathscr{A}$ one can find as good an approximation as one wants to any state in $\mathscr{T}_\pi$ in terms of states in $\mathscr{T}_{j(\pi)}$. Therefore in an *approximate* sense any faithful representation should be able to describe the basic physics (see section 3 for a discussion of this weak convergence in the context of measurement theory).

15. Motivated by this result, Haag and Kastler (1964) developed a general framework for a QFT in purely algebraic terms. Their assumptions were as follows:

(i) For every bounded open set in space-time B there is an abstract measure algebra $\mathscr{A}(B)$.

(ii) Isotony: If $B_1 \supset B_2$ then $\mathscr{A}(B_1) \supset \mathscr{A}(B_2)$

(iii) Microcausality: If B_1 and B_2 are completely space-like (no time-like path connecting B_1 to B_2) then $\mathscr{A}(B_1)$ and $\mathscr{A}(B_2)$ commute.

(iv) The closure of the set-theoretic union of all the $\mathscr{A}(B)$ is a C*-algebra, denote $\mathscr{A}$, called the *algebra of quasilocal observables*.

(v) Lorentz covariance: The inhomogeneous Lorentz group is represented by automorphisms $\alpha_L \colon \mathscr{A} \to \mathscr{A}$ such that $\alpha_L \mathscr{A}(B) = \mathscr{A}(LB)$ where LB is the image of the region B under the inhomogeneous Lorentz transformation L.

These axioms have been widely influential; a characteristic feature is that bounded functions of global observables do not themselves belong to the algebra—hence 'quasilocal algebra'; obviously the familiar unbounded observables such as energy and momentum cannot lie in a C*-algebra, but here not even bounded functions of the total energy, momentum, etc. can do so, for they pertain to the whole of space-time. Such observables can, however, be associated with the von Neumann algebra associated with $\mathscr{A}$. Again, let $\mathscr{D}$ be the set of all open bounded regions in space-time and let $\mathscr{A}(\tilde{D})$ be the quantum system generated by all $\mathscr{A}(B)$ with $B \in \mathscr{D}$ and $B \cap D = 0$. For any representation π of $\mathscr{A}$ the von Neumann algebra $\mathscr{L}_\pi = \cap_\mathscr{D} \pi(\mathscr{A}(\tilde{D}))''$ is called *the algebra of observables at infinity in the representation π*. This algebra was introduced by Lanford and Ruelle (1969) and has enjoyed extensive applications, particularly in thermodynamics and measurement theory (see below and section 3).

The relationship between the quasilocal algebra $\mathscr{A}$ and the fields is understood roughly as the relationship between a manifold and a coordinate system. The idea is that different sets of fields might be used to 'parameterize' the algebra corresponding to the different ways in which the local algebras are defined; a particular choice $B \to \mathscr{A}(B)$ for all B in $\mathscr{D}$ is called a 'local net', and the idea is that as B shrinks to a point the local algebras can be characterized by (it is hoped a finite set of) local fields. This point of view has arisen because there are in known cases many different fields which can be used to describe the same system and because the same set of fields (with different mass and coupling constants) can be used to describe different systems. Of course, given a set of fields one knows how to generate the quasilocal algebra (in this sense there exist models of the Haag and Kastler axioms), but the converse problem is unsolved. Some progress has been made in this direction, however (Fredenhagen and Hertel 1981).

There is another, more special, result which is relevant here. Let us assume that the $\mathscr{A}(B)$s are actually von Neumann algebras, and further (vi) Primitive causality: Let T be a time slice in $\mathscr{D}$ (of arbitrary time-like width t) and let B be causally dependent on T; then $\mathscr{A}(B) \subset \mathscr{A}(T)$ for any t (B_1 is causally dependent on B_2 if every ray in the past light cone of any point in B_1 intersects B_2). In a certain limit it is possible to show (Haag and Schroer 1961) that this axiom is independent of the remaining axioms (even including the spectrum condition that the energy-momentum generators have spectrum in the forward light cone), that there must exist a field equation which determines the field at any point in space-time from its expectation value on T, and that this field equation must have a hyperbolic propagation character. They gave no guide-lines on how to construct such a field, however.

16. Whilst this approach of Haag *et al.* was initially motivated by Fell's result, subsequent developments (guage theories, grand-unification, spontaneous symmetry breaking) in the ultra-relativistic theory have confirmed the physical significance of the correct choice of ground state and associated representation. One might therefore feel that the correct strategy is to find the 'true' physical vacuum, and the associated representation (a Fock representation) will then provide a 'true' concrete representation of dynamics. Since the Fock representation is unique up to unitary equivalence once the functional ϕ_F is fixed one could forget about the other representations. There are two arguments against this view; the first is that there are important applications of non-Fock representations in thermodynamics (see below); the second is that a Fock representation does not exist in curved space-time.

17. I refer to the extensive research programme, initiated by discoveries of Fulling (1973) concerning quantization in Rindler coordinates on Minkowski space, conducted through the 1970s and up to the present. On some interpretations (e.g. Davies 1984) these investigations undermine the objectivity of the particle concept even in Minkowski space-time—the particle content of a domain in space-time cannot be freed from the coordinate system used in quantization and is strictly dependent on both the motion of an observer and the particular structure of the particle detector which he employs. A less radical viewpoint (which enjoys wide concensus, see, e.g., Fulling (1984)) is that a particle definition is possible only for space-times with special symmetry properties, notably Minkowski or de Sitter space-times. The underlying problem is this: the particle concept, and with it the definition of the vacuum state, depends critically on the possibility of defining creation and annihilation operators, which is equivalent to performing an invariant and global decomposition of the field into its positive and negative frequency parts. This decomposition in turn requires the existence of a global time-like Killing vector field (the integral curves of which define isometries of the space-time) which is everywhere orthogonal to a family of space-like hypersurfaces. This is only possible when the space-time is *static* (Hawking and Ellis 1973). One can slightly relax this criterion to include stationary space-times (where these properties obtain at past and future infinity; see Ashtekar

and Magnon (1976)) and in this way obtain a particle definition in an asymptotic sense. For the Schwarzchild case the Hawking radition is just this: at the asymptotic approach of the Schwarzchild to the Minkowski space-time, the particle definition (fixed by the vacuum in the two geometries) differs by a thermal flux of particles 'originating' at the Schwarzchild radius. In general, however, one does not have even this much symmetry in realistic models of space-time.

One might hope to circumvent these problems in supersymmetric theories or, more generally, quantum gravity, but the usual view is that one must focus on purely local observables in the theory, which can be constructed as local functions of the fields (Fulling 1984). Certainly other axiom schemes such as the Garding–Wightman or the Osterwalder–Schrader axioms, do not appear flexible enough to accommodate this situation, for they place explicit demands from the outset of the existence and uniqueness of a globally invariant vacuum state.

18. The point here is not that no concrete model for the universe can be found; on the contrary, if a quantum system is truly universal, and if we can find 'the' state of the universe, we are assured of a concrete model via the GNS construction. If, further, dynamics can be expressed as an automorphism of the quantum system, then this concrete model may even host a unitary evolution and thus describe any physical phenomenon which actually obtains. The point is rather that in the absence of this ideal situation we have to idealize the universe in different ways appropriate to different phenomena, and this will require different representations. That is we may discover the 'ultimate' laws of the universe, but it is implausible that we will discover their unique concrete representation. (See also (3.15).)

19. I will conclude this section with a discussion of applicatons to (quantum) thermodynamic system. Of course, the same methods are applicable to classical statistical mechanics: for example the Wightman construction is in fact a variant of the GNS construction and can be directly applied to a classical system (to yield a Koopman system), with Wightman functions given by the n-point correlation functions. Incidentally, one obtains in this way a commuting operator 'field' (operator-valued measure) which can be thought of as the Koopman representation of a classical particle density observable; if one considers with surprise the appearance of a kind of 'field-many particle duality' in *classical* theory, one should also consider that classical thermodynamics is a perfectly good classical field theory.

In the quantum case the basic ideas are well illustrated in the Araki and Woods (1963) analysis of the infinite free Bose gas, one of the first applications of non-Fock representations to physics. These authors sought an infinite-volume finite density model; the first point is that obviously the density of a gas must be zero in any Fock representation for infinite volume, because Fock space contains only finite-particle states, and the number of particles for finite density must be infinite. Of course, the usual strategy is to calculate everything that one wants to know in a finite volume, and then let this volume tend to infinity (and the number of particles also) whilst keeping the density constant, to obtain the limiting therodynamic behaviour of the system. The key insight of Araki and Woods was that the density operator, if it exists, must be a constant in any given representation. In an irreducible representation it is easy to see this from a straightforward calculation of the infinite volume limit of the commutator of a bounded function of the density operator with an arbitrary element $W(f,g)$ of the Weyl algebra; this comutator vanishes in the limit, so that the density operator must lie in the commutant of the Weyl algebra. In particular, in an irreducible representation the commutant is trivial and the density operator must then be a multiple of the identity. In the Fock case, since this representation is irreducible, its value can be calculated on any state, in particular the vacuum, where of course its value is zero as expected.

It actually follows that the density must be a constant on any cyclic representation, with a value given by its expectation value on the cyclic vector. We now note that although the Fock representation does not contain infinite particle states we can still use this representation to define states of this character; one defines a finite density functional $\hat{\phi}_{F,V}$ for a finite-volume Fock representation, and then less V tend to infinity. The resulting functional is well defined and can be used to define a new representation via the GNS construction. As Araki and Woods showed, one thus obtains a representation for an infinite volume system at finite density. They also found that the density operator always lies in the centre of the resulting representations. This being so, it is a macroscopic (or classical) observable.

20. This strategy for defining macroscopic properties in the infinite-volume limit is quite general; one must start with the finite-volume system in the Fock representation, pick a state which is in some way typical of any finite sub-volume of the infinite system one wants to describe, allow the volume to go to infinity to arrive at a new

state (which will not lie in the original space of states), and then use this state to define a new representation. In general the states which one arrives at in this way are not pure, so that one ends up with reducible representations. One must expect this to be so, because the algebraic operations result in a change in the particle number—this is so even though the total particle number is not defined—but an infinite system cannot be affected by any finite change of particle number if it has non-zero density, and there must be some degeneracy present to reflect this fact.

21. Araki and Woods also extended their analysis to the finite temperature case with and without macroscopic occupation of the ground state; the general feature seems to be that intensive properties such as entropy density, chemical potential, temperature, and density are uniquely associated with a representation. Significant results have also been obtained in connection with the Bardeen, Cooper, and Schrieffer model of superconductivity (BCS model; see Schrieffer (1964), Haag (1962), and Jelinek (1968)) and for a class of Weiss models for ferro- and antiferromagnetism (Emch and Knopps 1970). In all these cases the states which generate these equilibrium representations satisfy the *KMS condition* (after Kubo (1957) and Martin and Schwinger (1959)), first formulated as a boundary condition on the analytic behaviour of thermal Green's functions. This condition arises naturally as an equilibrium conditions, as can be seen from the following argument: consider the Gibbs state for a finite volume system $\rho = e^{-\beta H}/\text{Tr}[e^{-\beta H}]$ at temperature $\beta = 1/kT$ with Hamiltonian H; under the automorphic time-evolution α_t one has
$$\langle \rho; \alpha_t(A)B \rangle = \frac{\text{Tr}[e^{-\beta H}e^{iHt}Ae^{-iHt}B]}{\text{Tr}[e^{-\beta H}]}$$
$$= \frac{\text{Tr}[e^{-\beta H}Be^{i(t+i\beta)H}Ae^{-i(t+i\beta)H}]}{\text{Tr}[e^{-\beta H}]}$$
$$= \langle \rho; B\alpha_{t+i\beta}(A) \rangle.$$

In particular for $B = I$ and analytically continuing to imaginary time, one sees that ρ must be time-invariant. Quite generally, any state ρ satisfying $\langle \rho;\alpha_t(A)B \rangle = \langle \rho;B\alpha_{t+i\beta}(A) \rangle$ is called a KMS state for the temperature β; this condition is called the KMS condition. It can be shown that a Gibbs ensemble of a finite system with temperature β and chemical potential μ for a finite system satisfies the KMS condition as the volume tends to infinity (with μ, β fixed) if the infinite volume limit of the local densities exists; the group of automorphisms with respect to which the generating state is KMS may be identified with the abelian Lie group of time translations and gauge transformations (one for each species of particle) (see Haag *et al.* 1967). On the other hand, any time-invariant state which is stable with respect to local perturbations is KMS with respect to such a symmetry group (Haag *et al.* 1974). The one exception is when the state is a ground state. This is a limiting case of a KMS state, and remarkably (Kastler and Takesaki 1979) the KMS condition there goes over to the positivity of the spectrum of the Hamiltonian (and the limiting KMS state becomes the vacuum).

22. There are numerous ideas which we have not touched upon: in particular the relationship with the theory of group representations, consequences of microcausality, superselection rules, and spontaneous symmetry breaking. However, none of these is as controversial or as philosophically important as the application of these techniques to measurement theory.

3. APPLICATIONS TO THE THEORY OF MEASUREMENT

1. From the preceding remarks classical properties can be associated with non-Fock representations (thermodynamic or collective representations). This fact opens the way to an approach to measurement theory on the basic philosophy that a measurement process always arises as a collective or thermodynamic phenomenon (and, more generally, that classical or directly observable phenomena always have this character). The simplest and most telling argument here is that such processes are always irreversible.

2. This emphasis, on the irreversible character of measurement, is equally a feature of the Copenhagen interpretation and its subjectivist elaborations; what distinguishes the present approach is that one is able to show that properties defined in the thermodynamic limit are truly classical in their probabilistic behaviour as a purely mathematical consequence of quantum theory.

3. The measurement problem is, however, specifically a realist issue. It will not be solved until one provides an analysis free of any critical dependence on unphysical idealizations. In the present context that means relaxing the thermodynamic limit: what then emerges is that there is a precise parallel between thermodynamic

behaviour and classical behaviour; both must themselves be understood as idealizations.

4. The first task, however, is to understand the positive contribution of the present approach to the measurement problem. To this end recall that the essential difficulty in measurement theory is to understand the transition $j(\lambda_+\psi_+^M + \lambda_-\psi_-^M) \to |\lambda_+|^2 j(\psi_+^M) + |\lambda_-|^2 j(\psi_-^M)$, $\lambda_\pm \in \mathbb{C}, |\lambda_+|^2 + |\lambda_-|^2 = 1$, that must surely be effected in a simple two-valued measurement M with final states $j(\psi_\pm^M)$ corresponding to the two possible experimental outcomes. The fundamental intuition is most succinctly expressed in the Schrödinger cat paradox: macroscopic properties are invariably subject to an ignorance interpretation of probability; that is to say, it must make sense to assume, at any instant, that they have some definite (if unknown) value. This interpretation of coherent superpositions is known to be untenable; from the analysis of propositional systems which admit coherent superpositions, we know that their lattice of propositions must be non-Boolean. We have therefore arrived at the classic measurement problem: how to effect the transition from coherent superposition to incoherent mixtures. A less widely appreciated problem also arises: how to show that the resulting mixtures determine a unique classical probability theory over the experimental outcomes (we shall come on to this presently).

5. The expectation value of some observable $A \in \mathscr{A}^M$ in the coherent superposition $j(\lambda_+\psi_+^M + \lambda_-\psi_-^M)$ differs from that in the incoherent mixture by cross-terms of the form $|(\lambda_+\psi_+^M, A\lambda_-\psi_-^M)|$. This leads to the idea that if all such cross-terms vanish for the *macroscopic* observables of a system, then there is no difficulty in interpreting a macroscopic state as a coherent superposition of states, because this description is *equivalent* to the description of the system as in an incoherent mixture with respect to all macroscopic observable properties.

6. This idea plays a central role in the Daneri, Prosperi, and Loinger (DPL) theory of measurement (Daneri *et al.* 1962), where the macroscopic observables are those which do not induce transitions between states belonging to distinct 'channels' of the measurement apparatus. These observables are typically time-averaged (Heisenberg picture) observables. A similar idea underlies Jauch's (1971) use of the partial trace to reduce pure states to mixtures.

7. In the present theory one can go much further, with no appeal to time-averages or conditionalization over the states of the microscopic

system. (3.5) expresses a 'minimal' requirement of a theory of measurement to make sense of the observable structure of events. It is a remarkable fact that in the algebraic approach not only do such cross-terms vanish for the classical algebra of observables, but for the whole algebra.

8. To understand this result, we need the notion of *disjointness*. Consider any faithful representation π of a C*-algebra $\mathscr{A}$, so that vectors ϕ_1, ϕ_2 in $\mathscr{H}_\pi$ determine pure states $j(\phi_1), j(\phi_2)$ in $\mathscr{T}$ and their associated GNS representations π_1, π_2 are irreducible. Obviously each of these representations is unitarily equivalent to some sub-representation of π (the restriction of π to some stable subspace in $\mathscr{H}_\pi$, namely that which contains ϕ_1, ϕ_2 respectively (see (2.5)). Will they be unitarily equivalent to each other? The answer is affirmative if for some $A \in \mathscr{A}$ $(\phi_1, \pi(A)\phi_2)_{\mathscr{H}_\pi} \neq 0$ (i.e. ϕ_1, ϕ_2 lie in the same sub-representation). In the general case, any two states w_1, w_2 in $\mathscr{T}$ are said to be *disjoint* when *no* sub-representation of π_{w_1} is unitarily equivalent to *any* sub-representation of π_{w_2}; an easy extension of the above reasoning shows that states w_1, w_2 are disjoint if and only if for every representation π such that there exists $\psi_1, \psi_2 \in \mathscr{H}_\pi$ with $w_i = j(\psi_i)$, $i = 1, 2$, and for every $A \in \mathscr{A}$, one has $(\psi_1, \pi(A)\psi_2)_{\mathscr{H}_\pi} = 0$. (Note that finite-dimensional Weyl systems possess no disjoint states by (2.10)).

9. As we have seen in (2.19) macroscopically distinguishable *irreducible* representations are unitarily inequivalent, hence their states are disjoint. The fact that such states differ in the expectation values of macroscopic observables alone is enough to ensure that *all* cross-terms vanish for every $A \in \mathscr{A}$, so that no observable whatever can distinguish their coherent superposition from their incoherent mixture. This is just what is needed to make sense of an ignorance interpretation of *microphysical* probabilities; a definite (if unknown) event occurs on the microscopic scale precisely when the component states which enter into a coherent superposition evolve into disjoint states (because of some coupling with a measurement instrument). Note that any classical event (whether or not a measurement) should be described in this way, so that quite generally it is in the transition from pure states to disjoint mixtures that the probabilistic nature of classical experience is to derive.

10. In this connection, a cautionary note: in general an incoherent mixtue of states does not admit an ignorance interpretation. This fact is intimately linked to the possibility of forming the coherent

superposition of states which enter into the convex sum of the mixed state. In this way the decomposition of a mixture into pure states (and hence its ignorance interpretation) is not unique. What is required is that, for an ignorance interpretation of a mixture to be possible, its component states must form a *simplex* (Choquet and Meyer 1963). The set of states, even pure states, on a C*-algebra is much too large. In view of (3.9), what is needed is that disjoint states form a simplex; there seems to be no general proof along these lines, although it has been assumed (Beltrametti and Cassinnelli 1981, p. 82).

11. In general we must deal with reducible representations. An important class of representations which are 'almost' irreducible are the *primary* representations (i.e. such that $\pi(\mathcal{A})''$ is irreducible). For a certain class of macroscopic observables (global densitites, i.e. intensive observables defined as special averages which lie in the algebra of observables at infinity), it is a simple theorem (Hepp 1972) that primary states that differ on a global density are disjoint. The importance of primary representations is this: they are associated with pure thermodynamic phases (Emch and Knopps 1970). In particular, *extremal* KMS states are primary. These are the extremals of the convex set of KMS states at a given temperature; their interpretation as pure thermodynamic phases was first proposed by Ruelle (1965), who also showed (Ruelle 1967; 1969) that these are the *only* equilibrium states which are dispersion-free on global densities (see also Sewell 1986, sect. 4.4 and appendix B for an extension of this result). As such they must yield distinct values to such observables and hence are also disjoint. Moreover, they form a *simplex* (Emch 1972, Th. 15, Ch. 2), and KMS states for different temperatures are also disjoint (Takesaki 1970).

12. These results are about as good as one could hope for; the KMS condition is defined relative to a symmetry group on the quantum system (2.21), and there are good reasons to believe this condition is ultimately a consequence of locality and an ergodic property of the symmetry group (Kastler and Takesaki 1982); in this sense the DPL theory of measurement is not so far removed from the present theory. Summarizing the DPL theory, Prosperi has declared:

In conclusion the possibility seems to exist of getting out of the paradoxes connected with the occurrence of interference terms among macroscopically distinguishable states, assuming that physical observables uncompatible with the macroscopic quantities or at least with some privileged set of such quantities do not exist. Since however the idea that every self-adjoint operator (apart from superselection rules) corresponds to an observable, at least in principle, is quite naturally built in the mathematical structure of quantum mechanics, a consistent and logically satisfactory introduction of such a principle should require some kind of reformulation of the theory and perhaps some deep change in it (Prosperi 1971).

It should be clear that the theory of quantum systems provides just such a reformulation of quantum mechanics and that, on the contrary, for systems of infinitely many degrees of freedom realizations of the algebra on some Hilbert space $\mathscr{H}$ appear quite naturally as *smaller* than $\mathscr{B}(\mathscr{H})$. One way of putting this situation is that macroscopic observables induce superselection rules on the representations of the algebra.

13. Our basic ontology is that all systems, macroscopic structures included, are quantum fields; as such we always have to deal with a system of infinitely many degrees of freedom even when we have finite particle number. However, what is actually needed for the above results is a class of non-Fock representations where the number operator is not defined. We must face the question of whether such representation can be interpreted realistically.

14. There is another problem, equally obvious, and in some ways even more intractable: KMS states, as all thermodynamic equilibrium states, are time-invariant. One can express this problem in an even more distressing manner as follows. Let $\alpha \in \text{Aut}(\mathscr{A})$ be an automorphism of a quantum system $\mathscr{A}$; for any state $\phi \in \mathscr{T}$ define the linear functional $\alpha^*(\phi): \mathscr{A} \to \mathbb{C}$ by $\langle \alpha^*(\phi); A \rangle = \langle \phi; \alpha(A) \rangle$, for all $A \in \mathscr{A}$. It is easy to show that $\alpha^*(\phi)$ is a state and that α^* is a bijective *affine* map $\alpha^*: \mathscr{T} \to \mathscr{T}$. Therefore pure states are preserved under any automorphic time-evolution. As in the usual notion of superselection sectors in RQFT, one can understand disjoint states as states which cannot be superposed to form pure states; therefore (or by direct proof) disjoint states are also preserved by automorphic time-evolution. This shows that there is no point in looking for non-Fock non-equilibrium states, because the mechanism which we have exploited to derive the transition from pure states to incoherent mixtures cannot be modelled as an automorphic evolution. Non-equilibrium representations cannot possess the desirable properties in (3.11).

15. One concludes that here, as in other applications, the thermodynamic limit is a tool for capturing certain features of the phenomen-

ology, whilst at the same time introducing unphysical characteristics which are not present in finite-volume models. It is instructive to review some typical features of the finite and infinite volume analysis in connection with thermodynamic phenomena; in a phenomenological sense, physical systems appear to have properties which are insensitive to the actual geometry of the system (intensive properties), phase transitions appear to be sharply defined with associated discontinuities in the thermodynamic potentials, distinct pure phases appear to coexist in equilibrium, and finally such systems appear to evolve irreversibly. None of these properties can be modelled within a finite-volume analysis: global averages of finite systems only converge to definite values as the number of degrees of freedom becomes infinite (Ruelle 1969), thermodynamic variables are continuous (indeed, analytic) for any finite system (a classical result; Lee and Yang (1949), Lebowitz (1968)), there exists a single, unique KMS state (hence extremal) for a finite system at given volume, temperature, and mass, and one can even show that metastable states (such as superheated or supercooled liquids) cannot be described in the standard finite theory (Fisher 1964). To this list it seems that one must add that a strict ignorance interpretation of probability can only be given in the infinite limit.

16. The analogy with Poincaré recurrences and the second law is helpful; in the classical theory a qualitative feature (reversibility and sensitivity to initial conditions) at the microscopic level cannot be eliminated in any finite model. Going to the thermodynamic limit one obtains irreversibility but then no truly dynamic behaviour can be described. Of course we know that this is more than an analogy—that essentially the same mathematical technique is being used here as to destroy coherence in the quantum theory. In a sense the thermodynamic limit has long provided a theory of measurement for statistical mechanics (relating microscopic properties to thermodynamic macroscopic ones); what is remarkable in the quantum case is not that the same representation theory yields a quantum thermodynamic description but that the space of states of this system should form a simplex.

17. The probabilistic behaviour of macroscopic systems therefore arises from the quantum indeterminism in the same way that irreversibility arises from a microscopic reversible system—as a theory of statics. How should we interpret this situation? The simplest idea is that the classical properties of the quantum system are to be

understod as *idealizations* of certain properties of macroscopic systems—that is, any weakly coupled system of at least (say) 10^4 massive degrees of freedom over a time-scale of the order of the thermal relaxation time at usual temperatures, (about 10^{-14} secs). Of course classical methods have been successfully used at much smaller scales than this, but our brief is the question of coherence, which is not a typical preoccupation of the chemist or crystallographer. These length scales (of order 10^{-8} m for dilute systems) are certainly sufficiently small to generate a classical field theory along the lines of Noll (1958), i.e. a continuum, fluid or thermodyanamics.

18. Unfortunately it is by no means clear that one can exclude the influence of systems past time-like or even spacelike to this volume in the measurement context; Einsteinian locality cannot be presumed in this field. (So much is surely the moral of the Aspect experiments.) Nevertheless we make the most simple assumption that the Cauchy problem is well posed on the intersection of a spacelike hypersurface with the past light-cone of a system of the indicated size, at least when taken in isolaton (an assumption that is still true of each separated system in EPR-type correlation). This leads to a considerable increase in the particle number of the 'effective' system (taken with its local environment determined in this way, of order 10^{-6} m and 10^{10} degrees of freedom).

19. The drift of these remarks is that a 'nice' decay of coherence with particle number would be 'sufficiently' rapid such that systems of this order of scale have 'approximately' classical properties. This raises the very serious problem of a residual coherence, however small and rapidly decaying, in the realm of macroscopic experience.

20. There are ways of evading this problem. The crude models which are available are non-relativistic so that questions of locality do not directly enter the theory; in the absence of any global definition of a vacuum in physically realistic space-times there is no compelling argument that the 'physical' representations must be of Fock type (cf. (2.17)). And there is a long and honourable tradition in physics of simply identifying the infinite limit with an empirical context in which the limit is palpably not reached. Scattering theory is entirely a theory about asymptotic limits and the problem of controlling these limits is rarely raised in the physics literature. It is not even necessary to consider infinite systems (Haag's theorem, cf. (2.13)) to see just how misleading our physical intuitions can be, or, to put it another way, how little they are respected by the mathematical models that we use.

One recalls the theorem (e.g. Thirring 1979, Th. 3.5.6) which says that no matter how small the perturbation H' of a Hermitian operator H_0, and even if H' is bounded or small relative to H_0, the spectrum of H' + H_0 will generally differ in type from that of H_0—and hence that in general no interaction picture exists (since the spectrum is invariant under unitary transformations).

21. But here, as in so much else, the measurement problem imposes its own harsh discipline. If the question 'when is a scattered particle free? has limited philosophical interest (there may not, after all, exist any truly free particle), the notion that some 'residual' coherence, however small, may contaminate the description of macroscopic events, has profound significance for our notions of macroscopic realism.

22. In developing models of the measurement process which make use of infinite systems it is crucial that one obtains estimates on large but finite systems of which we suppose the infinite systems are idealizations. Failing this, at least let us be clear as to what length-scale is supposed to be identified with the infinite volume limit. One might summarize the possible interpretations by means of the comparison of the several length-scales which follow from general considerations—the cosmological, local environment, small-macroscopic, and microscopic—with the *two* qualitative distinctions available in the present theory—the local and global. It is not even that the ambiguity concerns only the reference of the global properties of the system. There is a long tradition, particularly in the more mathematical literature, of interpreting the local observables (or algebra of observables) with possible experiments that can be performed in the associated space-time region (cf. (1.9)). This is surely of local environment or small-macroscopic length scale.

23. With so many problems besetting the use of infinite systems one might wonder whether after all we may not resign ourselves to dealing with the measurement problem with the more elementary techniques of finite systems. I feel the strongest counter-argument is this: solving the measurement problem on a realist basis must at the same time provide a description of the approach to equilibrium of many-particle systems, since that is what actually occurs in any physical experiment (and in almost all macroscopic phenomena). One might be optimistic in a fringe activity like measurement theory of finding a simple, finite resolution of the problem, but in statistical (quantum and classical) mechanics one has a major part of modern

physics. (Incidentally, it is by no means coincidental that in one popular example of a 'simple, finite resolution', namely the Everett 'many worlds' interpretation, one is inexorably led to the question of how to define the ensemble of worlds which identify the world of our experience, as it appears to evolve, from the class of all worlds. This is statistical mechanics in disguise.)

A more physical response is that one cannot formulate the dynamical approach to equilibrium in any *asymptotic* sense of a closed and finite system. That is because as time evolves, more and more particles enter the causal environment of the system studied and the effective particle number must increase as the cube of the time. Only in the infinite limit can this change in particle content leave the representation invariant and permit the use of asymptotic time limits.

24. Pursuing this line one might think that the proper approach is to consider finite *open* systems, in particular that the evolution of such a system which takes account of the influence of the causal environment should be a contraction semi-group (Davies 1976). That it must be non-automorphic is clear. There is, further, a general theory due to Lewis and Thomas (1975), which shows us how to construct an (infinite) classical environment and an automorphic evolution acting on the finite system embedded in this environment which 'mimics' the original evolution. Massen (1982) and (1984) and Hannabus (1984) have provided quantized versions of the Lewis and Thomas dilation theory (in the Bose and Fermion case respectively); the latter, in particular, has shown that in the limit of an underdamped system the combined state of the system and environment become disjoint in the infinite time limit.

There are obvious attractions to this approach; there is a qualitative distinction available between the finite measuring apparatus and its infinite environment, and one can choose to consider the infinite environment as a convenient mathematical artifice for obtaining asymptotic estimates on the 'real' finite system. However it remains unclear how the disjointness obtained in the limit is to be understood in terms of this finite system. Neither have estimates been obtained on the asymptotic limit. For all that, the synthesis of ideas contained in this theory of measurement is an impressive achievement.

25. The first algebraic theory of measurement tried rather to confront the original difficulty (3.14) head on; this is the theory of

The Algebraic Approach to Quantum Field Theory

Klaus Hepp (1972), apparently an elaboration of ideas of M. Fierz and R. Jost. The idea is to find an infinite system $\mathscr{A}$ and a group of automorphisms α_i ($i \in \mathbb{Z}$), which evolve a state w 'close to' some disjoint state w'. In this limit we know that no automorphism can actually map w onto w'; therefore for any i there will be some element $A \in \mathscr{A}$ such that $\langle \alpha_i^*(w) - w'; A \rangle \neq 0$. Can we be sure that, whilst non-zero, such a cross-term can always be made small? The answer is negative: if this were so, i.e. for any ε we can find a number $N(\varepsilon)$ such that $\langle \alpha_n^*(w) - w'; A \rangle \leq \varepsilon$ for $n \geq N(\varepsilon)$ and for any $A \in \mathscr{A}$, the sequence $\alpha_n^*(w)$ would be strongly convergent on the (disjoint) state w'. This means that, in any representation π in which $\alpha_n^*(w) \in \mathscr{T}_{j(\pi)}$, then so too would $w' \in \mathscr{T}_{j(\pi)}$ (since this set is norm-closed by definition) so that w' could not be disjoint, a contradiction. Recalling, however, that $\mathscr{T}_\pi$ is the weak * closure of $\mathscr{T}_{j(\pi)}$ (cf. (2.14); every state in $\mathscr{T}_\pi$ can be approximated by a net of vector states in $\mathscr{T}_{j(\pi)}$) one might hope to find a sequence $\{\alpha_n^* w\}$ which is weak * convergent on w', i.e. that for any ε and any *finite* set $\mathscr{M} \subset \mathscr{A}$ there is a number $N(\varepsilon, \mathscr{M})$ such that $\langle \alpha_n^*(w) - w'; A \rangle \leq \varepsilon$ for $n \geq N(\varepsilon, \mathscr{M})$ and for all $A \in \mathscr{M}$.

26. It is not too surprising that if $w_{i,n}$, $i = 1, 2$ are two such sequences, weakly converging on two disjoint states w_i, then all cross-terms in any representation converge weakly to zero. More precisely, let π_n be a sequence of representations of $\mathscr{A}$ with $w_{i,n} = j_{\pi_n}(\phi_{i,n})$, $\phi_{i,n}$ vectors in $\mathscr{H}_{\pi_n}$; then for any ε, $\mathscr{M}$, there is an $N(\varepsilon, \mathscr{M})$ with $(\phi_{1,n}, \pi_n(A)\phi_{2,n})_{\mathscr{H}_{\pi_n}} \leq \varepsilon$ for all $n \geq N(\varepsilon, \mathscr{M})$ and $A \in \mathscr{M}$ (Hepp 1972, Lemma 3).

27. This result has a natural positivist interpretation (cf. (2.14)): that for any finite set of measurements (and the number of observations that have been made or ever will be made is finite) the coherence terms vanish as $t \to \infty$. Repudiating this positivist philosophy, Bell has remarked 'while for any given observable one can find a time for which the unwanted interference is as small as you like, for any given time one can find an observable for which it is as big as you do *not* like' (Bell 1975). All of the automorphic models constructed by Hepp lead to asymptotic disjointness; the analogy, that 'the introduction of an asymptotic condition into measurement theory is as natural as elsewhere in microphysics, where S-matrix theory is sometimes considered as the ultimate receptacle of all physics' (Hepp 1972) has already been considered in (3.20). Incidently, it should be clear that there is no general obstacle to the approach to disjointness

in finite times if one is prepared to consider non-automorphic evolutions. For closed systems, however, this appears unacceptable for a variety of reasons (for lack of space we shall not attempt to deal with this issue here).

28. There is another issue at stake, and that is the (usually implicit) shift in emphasis from the classical to the local observables as the proper objects of macroscopic experience (cf. (3.22)); Bell makes this shift explicitly, although he offers no justification for it. No doubt this explains why he offers no realist critique on the use of infinite systems. Yet this question is fundamental: it is, after all, a bonus that we should obtain a strict ignorance interpretation of microphysical probability, (i.e. lack of coherence between states which describe different experimental outcomes with respect to all microscopic observables); this is even more than the projection postulate. Would such residual coherence be observable? Of course we use data on classical observables to deduce the expectation values of *some* microscopic observables; but need *these* be subject to a residual coherence? And what of the much more pressing question: When do the *classical* properties of a measurement device change?

29. These ambiguities weaken a widely held interpretation of the various models that have been proposed (Frigerio 1974; Whitten-Wolfe and Emch 1976; Hannabus 1984; Zurek 1981) that the coherence is never really lost, but propagates out over vast regions of space where it is no longer observable. The difficulty is that the classical properties are globally defined, typically as the algebra of observables at infinity. One wants simultaneously to asert that the global observables describe small macroscopic systems, yet that there exist local observables which describe properties pertaining to 'vast' regions of space.

One might conclude that the classical observables of a system play a merely technical role in the present approach to measurement theory. Yet this is to abandon the fundamental intuition that classical phenomenology—both with respect to probabilistic behaviour and to thermodynamic behaviour—arises as collective phenomena involving large numbers of elementary (or effectively elementary) systems. There is surely something inconsistent with the view that observed thermodynamic properties be associated with classical observables in one context of statistical mechanics yet be relegated to a purely technical role in measurement theory.

30. Considering this difficulty, one feels a certain exasperation. The

association of large length-scales, on the one hand with classical properties, on the other hand with an unobservable residual coherence, is surely too simplistic on both scores. It is, surely, rather a matter of the 'kind' of observable: whether approximating a quasi-equilibrium collective property, or an arbitrary unstable microscopic property, irrespective of length-scale. And it is in the latter that we might hope to find 'residual coherence', possibly at any finite time. It is even possible that coherence will vanish in finite times for a restricted class of observables which approximate classical observables.

31. This possibility is surely susceptible to mathematical analysis; the obvious step is to look for comparative estimates on the decay of coherence on the two classes of observable. In this respect models such as those of Hannabus (1984) and Frigerio (1974), which explicitly distinguish a finite sub-system from an infinite system, may be useful if interpreted somewhat differently: that the finite system be associated with a class of far-from-equilibrium properties.

32. But what if some degree of residual coherence is indeed invasive at the macroscopic level? We hencefoward consider this question, under the reasonable assumption that nevertheless there will be an extremely rapid fall-off in coherence for macroscopic observables which describe quasi-equilibrium properties of macroscopic systems. One can put the situation as follows: that whatever sort of a thing macroscopic quantum coherence (MQC) might be, it is the sort of thing which falls off rapidly with time and complexity of the effective system. There is therefore a very strong disturbance effect of measurement; the coupling of one experiment instrument to another enormously increases the complexity of the total effective system. In this way the von Neumann infinite regress which appears to follow from elementary quantum theory is rapidly damped.

33. Some phenomenological consideration: the *subjective* experience of radiation is a very sensitive measurement device (observers have been known to detect light levels involving only a few quanta). The effective system that we must consider has however a very large number of degrees of freedom (of the order of the human brain, say 10^{28}). If the fall-off in coherence is fast enough for a system of this scale the experience of MQC (due to *this* cause) will be of negligible duration in comparison to the time-scale of subjective experience. This will apply *a fortiori* to the *apparent* behaviour of laboratory instruments. Nevertheless there are indirect methods; in particular it

may be possible to exploit the Bell inequality as a test for MQC in the behaviour of certain types of macroscopic systems. Consider a macroscopic system with some property Q which, on measurement, is always found to have one of two values ± 1. Fix a time interval τ and time t and define the observables Q^t_{ij}, i, $j = 0$, 1, 2, $Q^t_{ij} = Q(t+i\tau) Q(t+j\tau)$. Now prepare an ensemble of such systems in a similar way at t_0 and define the ensemble average $\langle Q^t_{ij} \rangle$; if now we assume that the measurement of $Q(t+i\tau)$ does not effect the value of $Q(t+j\tau)$ (whether measured or not, this is equivalent to the locality assumption in standard applications of the Bell inequality), then the inequality $|\langle Q^t_{13}\rangle + \langle Q^t_{23}\rangle + \langle Q^t_{14}\rangle - \langle Q_{24}\rangle| \leq 2$ will hold if $Q(t+i\tau)$ always has a definite value whether or not it is measured (see, e.g., Clauser & Shimony (1978)). The Bell inequality therefore provides, under the stated conditions, a phenomenological tool for discovering MQC. The theoretical problem of selecting a 'suitable' class of systems, and in particular justifying the condition on the independence of the $Q(t+j\tau)$ and on the choice of $t-t_0$, τ, is, however, somewhat daunting. We refer to Leggett (1985) for a proposal in which the magnetic flux trapped in one of two potential minima of a SQUID (superconducting quantum interference device) and oscillating between them via quantum tunelling plays the role of the observable Q above. This model is, however, based on formal considerations only; neither is it clear that the magnetic flux should, in this (still speculative) phenomenon, be considered a macroscopic system.

34. On estimates: Frigerio (1974) considered a finite apparatus coupled to an infinite environment as a finite volume of a one-dimensional spin-$\frac{1}{2}$ lattice. In this model local observables were again considered as measurable (3.28): the estimate t^{1-n} was obtained (with n the number of lattice sites) for the ratio of the interference terms to the difference of the expectation values of such observables in a suitable class of states. Obviously the most pressing need is to produce estimates for more realistic models; for those observables which actually describe our macroscopic experience we may expect the fall-off to be so rapid as to be effectively instantaneous. If successful in this task, one will have the curious situation that these effects will occur *whether or not* they are considered relevant to the measuring process, as remarked by Hannabus (1984).

35. We do not yet have a satisfactory measurement theory. But the fundamental ideas seem to be all in place. Is it physics, or is it mathematics? I will conclude with a curiosity, a fragment of a

personal memo written by a physicist, at about the same time that Segal was perfecting the representation theory of C*-algebras (compare with (3.23) above):

When you start out to measure the property of one (or more) atom, say, you get, for example, a spot on a photographic plate which you then interpret. But such a spot is really only more atoms and so in looking at the spot you are again measuring the properties of atoms, only now it is more atoms. What can we expect to end with if we say can't see many things about one atom precisely, what in fact can we see? Proposal: *Only those properties of a single atom can be measured which can be correlated (with finite probability) (by various experimental arrangements) with an unlimited number of atoms.* R. Feynman, private memo, 1946 (emphasis mine; quoted in Schweber 1986, p. 463).

REFERENCES

ALBERT, A. A. (1934): 'On a Certain Algebra of Quantum Mechanics', *Ann. Math.* **35**, 65–73.

ARAKI, H., and WOODS, E. (1963): 'Representations of the Canonical Commutation on Relationships Describing a Non-Relativistic Free Bose Gas', *J. Math. Phys.* **4**, 637–62.

ASHTEKAR, A., and MAGNON, A. (1975): 'Quantum Fields in Curved Space-Times', *Proc. Roy. Soc. (Lond.)* **A 346**, 375–94.

BELL, J. (1975): 'On Wave Packet Reduction in the Coleman–Hepp Model', *Helv. Phys. Acta* **48**, 93–8.

BELTRAMETTI, E., and CASSINELLI, G. (1981): *The Logic of Quantum Mechanics* (Reading, Mass.: Addison-Wesley).

CHOQUET, G., and MEYER, P. (1963): 'Existence et Unicité des Représentations Intégrales dans les Convexes Compacts Quelconques', *Ann. Inst. Fourier* (Grenoble) **13**, 139–54.

DANERI, A., LOINGER, A., and PROSPERI, G. (1962): 'Quantum Theory of Measurement and Ergodicity Conditions', *Nucl. Phys.* **33**, 297–319.

CLAUSER, J. F., and SHIMONY, A. (1978): 'Bell's Theorem: Experimental Tests and Implications', *Rep. Prog. Phys.* **41**, 1881.

DAVIES, B. (1976): *Quantum Theory of Open Systems* (London: Academic Press).

DAVIES, P. (1984): 'Particles Do Not Exist' in *Quantum Theory of Gravity*, ed. S. Christensen (Bristol: Adam Hilger).

DIXMIER, J. (1957): *Les C*-Algèbres et Leurs Representations* (Paris: Gathier-Villars).

EMCH, G. (1972): *Algebraic Methods in Statistical Mechanics and Quantum Field Theory* (New York: John Wiley).

——and KNOPPS, H. (1970): 'Pure Thermodynamic Phases as Extremal KMS States', *J. Math. Phys.* **11**, 3008–18.

FELL, J. (1950): 'The Dual Spaces of C*-Algebras', *Trans. Am. Math. Soc.* **94**, 365–403.

FISHER, M. (1964): 'The Free Energy of a Macroscopic System', *Arch. Rat. Mech. Annal.* **17**, 377–410.

FREDENHAGEN, K., and HERTEL, J. (1981): 'Local Algebras of Observables and Pointlike Localized Fields', *Comm. Math. Phys.* **80**, 555–61.

FRIGERIO, A. (1976): 'Quasi-Local Observables and the Problem of Measurement in Quantum Mechanics', *Ann. Inst. H. Poincaré* **A21**, 259–70.

FULLING, S. (1973): 'Non-uniqueness of Canonical Field Quantization in Riemannian Space-Time', *Phys. Rev.* **D7**, 2850–62.

—— (1983): In *Gauge Theory and Gravitation: Lecture Notes in Physics 176* (Berlin: Springer-Verlag), 101–6.

GELFAND, I., and NAIMARK, M. (1943): 'On the Imbedding of Normed Rings into the Ring of Operators in Hilbert Space', *Mat. Sborn* NS **12**, 197–213.

GLEASON, A. (1957): 'Measures on the Closed Subspaces of a Hilbert Space', *J. Math. and Mech.* **6**, 885–93.

HAAG, R. (1962): 'The Mathematical Structure of the Bardeen–Cooper–Schrieffer Model', *Nuovo Cimento* **25**, 287–98.

—— HUGENHOLTZ, N., and WINNINK, M. (1967): 'On the Equilibrium States in Quantum Statistical Mechanics', *Comm. Math. Phys.* **5**, 215.

—— and KASTLER, D. (1964): 'An Algebraic Approach to Quantum Field Theory', *J. Math. Phys.* **5**, 848–61.

—— —— and TRYCH-POHLMEYER, E. (1974): 'Stability and Equilibrium States', *Comm. Math. Phys.* **38**, 173.

—— and SCHROER, B. (1962): 'Postulates of Quantum Field Theory', *J. Math. Phys.* **3**, 248–56.

HANNABUS, K. (1984): 'Dilations of a Quantum Measurement', *Helv. Phys. Acta* **57**, 610–20.

HAWKING, S., and ELLIS, G. (1973): *The Large-Scale Structure of Spacetime* (Cambridge: CUP).

HEPP, K. (1972): 'Quantum Theory of Measurement and Macroscopic Observables', *Helv. Phys. Acta* **45**, 237–48.

JAUCH, J. (1971): 'Foundations of Quantum Mechanics' in B. d'Espagnat ed.,), *Proc. Int. School of Physics 'Enrico Fermi'* **49**, (New York: Academic Press).

JELINEK, F. (1968): 'BCS-Spin Model, Its Thermodynamic Representations and Automorphisms', *Comm. Math. Phys.* **9**, 169–75.

JORDAN, P. (1933): 'Uber die Multiplikation quantenmechanischer Grossen', *Zeit. f. Phys.* **80**, 285–91.

—— VON NEUMANN, J., and WIGNER, E. (1934): 'On an Algebraic Generalization of the Quantum Mechanical Formalism', *Ann. Math.* **35**, 29–64.

KADISON, R. (1967): 'Lectures on Operator Algebras' in *Cargese Lectures in Theoretical Physics*, ed. F. Lurcat (New York: Gordon & Breach), 41–87.

KAKATUNI, S., and MACKEY, G. (1944): 'Two Characterizations of Real Hilbert Space', *Ann. Math.* **45**, 50–8.

KASTLER, D., and TAKESAKI, M. (1979); 'Group Duality and the Kubo–Martin–Schwinger Condition', *Comm. Math. Phys.* **70**, 193–212.

—— —— (1982): 'Group Duality and the Kubo–Martin–Schwinger Condition', *Comm. Math. Phys.* **85**, 155–76.

KOCHEN, S., and SPECKER, E. (1967): 'The Problem of Hidden Variables in Quantum Mechanics', *J. Math. and Mech.* **17**, 59–87.

KUBO, R. (1957): 'Statistical Mechanical Theory of Irreversible Processess', *J. Phys. Soc. (Japan)* **12**, 570–86.

LANFORD, O., and RUELLE, D. (1969): 'Observables at Infinity and States with Short-Range Correlations in Statistical Mechanics', *Comm. Math. Phys.* **13**, 194–215.

LEBOWITZ, J. (1968): 'Statistical Mechanics—A Review of Selected Rigorous Results', *Ann. Rev. Phys. Chem.* **19**, 389–418.

LEE, T., and YANG, C. (1952): 'Statistical Theory of Equations of State and Phase Transitions: II Lattice Gas and Ising Model', *Phys. Rev.* **87**, 401–19.

LEGGETT, A. J. (1985): 'Quantum Mechanics versus Macroscopic Realism: Is there Flux there when Nobody Looks?', *Phys. Rev. Lett.* **54**, 857.

LEWIS, D., and THOMAS, L. (1975): 'On the Existence of a Class of Stationary Quantum Stochastic Processes', *Ann. Inst. H. Poincaré* **A22**, 241–8.

MACKEY, G. (1951): 'Induced Representations of Groups', *Am. J. Math.* **73**, 576.

—— (1968): *Induced Representations and Quantum Mechanics* (Reading, Mass: Benjamin).

MARTIN, P., and SCHWINGER, J. (1959): 'Theory of Many-Particle Systems: I', *Phys. Rev.* **115**, 1342–73.

MASSEN, H. (1982): 'A Quantum Field Acting as a Heat Bath', *Phys. Lett.* **91A**, 107–11.

—— (1984): 'Return to Thermal Equilibrium by the Solution of a Quantum Langevin Equation', *J. Stat. Phys.* **34**, 239–61.

NOLL, W. (1958): 'A Mathematical Theory of the Mechanical Behaviour of Continuous Media', *Arch. Rat. Mech. Anal.* **2**, 197–226.

PLYMEN, R. (1968a): 'C^*-algebras and Mackey's Axioms', *Comm. Math. Phys.* **8**, 132–46.

—— (1968b): 'A Modification of Piron's Axioms', *Helv. Phys. Acta* **41**, 69–74.

PROSPERI, G. (1971): 'Macroscopic Physics and the Problem of Measurement in Quantum Mechanics' in B. d'Espagnat (ed.), *Proc. Int. School of Physics 'Enrico Fermi'* **49**, (New York: Academic Press), 97–124.

RUELLE, D. (1965): 'Correlation Functionals', *J. Math. Phys.* **6**, 201–20.

—— (1967): 'A Variational Formulation of Equilibrium Statistical Mechanics and the Gibbs Phase Rule', *Comm. Math. Phys.* **5**, 324.

—— (1969): *Statistical Mechanics* (Reading, Mass.: Benjamin).

SCHRIEFFER, J. (1964): *Theory of Superconductivity* (Reading, Mass.: Benjamin).

SCHWEBER, S. (1986): 'Feynman and the Visualization of Space-Time Processes', *Rev. Mod. Phys.* **58,** 449–508.

SEGAL, I. (1947): 'Postulates for General Quantum Mechanics', *Ann. Math.* **48,** 930–48.

—— (1967): 'Representations of the Canonical Commutation Relationship' in *Cargese Lectures in Theoretical Physics*, ed. F. Lurcat (New York: Gordon & Breach), 107–70.

SEWELL, G. (1986): *Quantum Theory of Collective Phenomena* (Oxford: Claredon Press).

STUECKELBERG, E., and GUENIN, M. (1962): 'Quantum Theory in Real Hilbert Space', *Helv. Phys. Acta* **35,** 673–95.

TAKESAKI, M. (1970): 'Disjointness of the KMS State of Different Temperatures', *Comm. Math. Phys.* **17,** 33–41.

—— (1979): *Operator Algebras* The Theory of Vol. 1 (New York: Springer-Verlag).

THIRRING, W. (1979): *A Course in Mathematical Physics 3: Quantum Mechanics of Atoms and Molecules* (Berlin: Springer-Verlag).

VARADARAJAN, V. (1968): *Geometry of Quantum Theory*, Vol. 1. (New York: Van Nostrand).

VON NEUMANN, J. (1936): 'On an Algebraic Generalization of the Quantum Mechanical Formalism I', *Mat. Sborn.* **1,** 415–84.

WHITTEN-WOLFE, B., and EMCH, G. (1976): 'A Mechanical Quantum Measuring Process', *Helv. Phys. Acta* **49,** 45–55.

ZUREK, W. (1981): 'Pointer Basis of Quantum Mechanics: Into What Mixture Does the Wave Packet Collapse?', *Phys. Rev'*. **D24,** 1516–25.

INDEX

abstract quantum system 149
affordance 60, 65–9
Aharanov–Bohm effect 57–8, 127
algebra
  C*-algebra 149–83
  Jordan algebra 150–1
    special 151
  of observables at infinity 165
  r-number algebra 150–1
  R*-algebra 154
  Segal algebra 152
    special 154
  von Neumann algebra 159
  Weyl algebra 162
anomalies 130, 141
anthropic principle 26, 32
asymptotic freedom 76
asymptotic completeness 19
axiomatic method 141, 144–8
  see also quantum field, Haag Kastler axioms

bare mass 75
Bell, J. 179
Bohr, N. 65–71
  atomic model 26
  complementarity 29–30
Born, M. 149

canonical commutation relationships 161–162
  equal time 101–2, 108
  for non-local fields 103–4
  see also algebra, Weyl
Cartwright, N. 60, 65
causality 65
  and classical electrodynamics 142
  and micro-causality 27–8, 34–5
CCR's see canonical commutation relationships
classical statistical mechanics 11, 167, 175
  see also representation, thermodynamic; states, KMS; thermodynamic limit
classical field theory 167, 176
classical limit 16

coherence 171–4
  macroscopic quantum coherence 181
  residual coherence 176–8
  SQUID 182
  see also states, disjoint; superposition
commutant, of a concrete C*-algebra 159
complementarity 29, 66
  see also Bohr, N.
creation and annihilation processes 11, 26, 43–5
  hyperplane dependent 104

Dirac, P. 17, 26, 137, 150
Dirac equation
  and hole theory 141
  and locality 140
  and primitive causality 140
  and second quantization 141

Eddington, A. 125
effective Lagrangian 54
Einstein, A. 29

Feynman, R. 183
Feynman diagrams 18, 47–52
  and experiment 62
  and visual imagery 60, 62–5
  see also perturbation theory
fibre bundle 124, 127
  see also gauge theory, geometrical interpretation
Fine, A. 27, 29
Fock space see particle representation; representation, Fock
force 16

gauge
  potential 124
  principle 119, 122
gauge theory 54–7, 117–32
  connections with general relativity 120
  geometrical interpretation of 124–32
  and isospin 126
  origins of 118–20
  and renormalization 118
  supergravity 123, 128
  see also quantum field

geometry
  and the geometrical programme 117
  geometrization of physics 117, 124–32
  physical vs. mathematical 124–32
ghost field *see* quantum field, ghost
Gibson, J. 65–7
Gleason's theorem 159–60
GNS construction 157

Haag, R. 163
harmonic oscillator interpretation 13, 17, 33
Heisenberg, W. 150
hyperplane, space-like 93–5
  dependence 96–7

ignorance interpretation of probability 172–3
impact parameter 76
individuality 10
  spatio-temporal 10
  transcendental 10
isospin 121–2, 126

Jordan, P. 149

Kaluza–Klein theory 16, 126, 129
Kant, E. 66
KG equation
  indefinite norm 139
  and negative energy states *see* states, negative energy
  non-local form 138
  positive definite norm 140
  and second quantization 141
Klein paradox 140–1
KMS condition 169
Koopman system 156–7
Kuhn, T. 28

Landau singularity 35, 83–4
lattice approach to quantum theory 151
locality
  Newton–Wigner 97
  hyperplane dependence 97
  and relativistic wave equations 138
  see also microcausality
Lorentz group, inhomogeneous
  Casimir invariant 105
  and gauge theory 121
  generators of 100–2
  and geometry 121, 125

  as local symmetry 124
  unitary representation of 100

measure assignment 152
measurement process 46, 153, 170–83
  Daneri–Prosperi–Loinger theory 171, 173–4
  Everett–Wheeler many worlds interpretation 178
  Hepp–Fierz–Jost theory 179–80
microcausality 27, 141, 143, 164
Miller, A. 60
monopole, t'Hooft–Polyakov 22, 126

Newton, I. 10 n., 29–30
number operator 13, 17

ontology 29–31

particle
  elementary 144
  as field excitation 15, 18
  and field underdetermination 10, 17, 167
  intermediate vector 62, 68–9
  interpretation, extended 17–18, 21
  representation 14, 17, 21
  see also representation, Fock
  virtual 9, 19, 43–58, 59–60
  as affordance 68
  and realist interpretation *see* realism, virtual particle
  and wave duality 9, 17, 26
perturbation theory 19–20, 44–56
  see also Feynman diagrams
Piron representation theorem 154
Planck, M. 28, 30–1
position operator, in relativistic theory 143
primitive causality 140–1, 165
probability measure 159–60
propagators 19–20, 48–54
  and imagery 64

quantization
  equivalence of 15–18
    see also particle, and field underdetermination
  field quantization 13–14
  path integral 47–53
  second quantization 14–15, 43–5

Index

quantum field
  asymptotically free 164
  Bose gas 168
  boson–fermion distinction 16
  complex scalar 15
  electroweak theory 33, 35
  Faddeev–Popov ghost field 54–7
  Fermi theory of weak interactions 35
  Haag–Kastler axioms 164–5
  Haag scattering theory 145, 147
  hyperplane–dependent 103–15
  interpretation, in comparison to quantum mechanics 25–8
  and local couplings 143
  and methodology 34, 60–1
  QCD 35, 78, 146
  QED 52, 74, 83–4, 143, 144, 146
  realist interpretation of 27, 152–5
  real scalar 13
  self–interacting, scalar 49, 102, 145–6
  and S-matrix 145
  and triviality 146
  Wightman theory 145, 146
quantum logic 159–60, 171
quark confinement 32, 78

realism
  and classical theory 29
  and Feynman diagrams 61
  and quantum field theory see quantum field theory, realism
  and virtual particles 65, 68–71
Redhead, M. 25
renormalization 18, 31, 73–88, 163–4
  actual infinities interpretation 85–6
  charge renormalization 79–84
  cut-offs approach 86
  finite corrections 76–9
  and finite theories 88, 99–100
  mask of ignorance approach 87–8
  mass renormalization 74–9
  and non-locality 98
  and the Planck length 99
  renormalization group 79
  with cut-off 75
representation
  of a C^*-algebra 156
  faithful 158
  Fock 163
  irreducible 158
  Schrodinger, of the CCR's 161
  thermodynamic 168–70, 174–83
  see also state, KMS

universal 160
unitarily inequivalent 158
weak equivalence 164
running coupling constants 78

Saunders, S. 71 n.
Schrodinger's cat 171
  see also measurement theory
Segal, I. 151
S-matrix theory 18, 33, 34, 35, 36, 147–8
spectral theorem 155–6
spin-statistics theorem 56
states
  asymptotic 21
  bound 21
  collision 18
  on a C^*-algebra 152–70
  cyclic vector, of a representation 156
  disjoint 172
  dispersion-free 153–4, 156
  generating, of a representation 156–7
  KMS 169–70
  negative energy 138
  normal 160
  primary 173
  soliton 22
  vector state, of a representation 158
Stein, H. 10 n.
Stone's theorem 162
superposition
  as central problem of interpretation 27, 33–4
  of macroscopic systems 27
  of states of distinct particle number 17, 26–7
  see also coherence
superstring theory 32, 35, 99, 123, 129–32
symmetry
  internal 121, 124
  invariance and covariance 120–1
  general covariance 122–3
  global and local 121, 126
  of states 12
  supersymmetry 16, 123, 127–9
  transformations, active and passive 16

Teller, P. 17, 26, 31
time operator 142–3
thermodynamic limit 168–9, 175–83
  see also classical, statistical mechanics
two-slit experiment 53

vacuum 17, 59, 69–70
  expectation values 48
  fluctuations 18
value assignment 152
vector potential 57–8
von Neumann, J. 150, 151
von Neumann uniqueness theorem 162

Weinberg, S. 26, 33, 34
Weingard, R. 17

Weyl, H. 119–20
Wightman, A. 145
Wigner, E. 144, 151
Wilson, K. 146

Yang, C. 126–7

zero-point energy 13, 17
  see also vacuum, fluctuations